Lecture Notes in Statistics

Edited by D. Brillinger, S. Fienberg, J. Gani, J. Hartigan, and K. Krickeberg

34

Douglas E. Critchlow

Metric Methods for Analyzing Partially Ranked Data

Springer-Verlag
Berlin Heidelberg New York Tokyo

Author

Douglas E. Critchlow
Department of Statistics, Purdue University
West Lafayette, Indiana 47907, USA

Mathematics Subject Classification (1980): 62A05, 62F07

ISBN 3-540-96288-3 Springer-Verlag Berlin Heidelberg New York Tokyo
ISBN 0-387-96288-3 Springer-Verlag New York Heidelberg Berlin Tokyo

Library of Congress Cataloging-in-Publication Data. Critchlow, Douglas Edward. Metric methods for analyzing partially ranked data. (Lecture notes in statistics; 34) Bibliography: p. Includes index. 1. Ranking and selection (Statistics) 2. Metric spaces. I. Title. II. Series: Lecture notes in statistics (Springer-Verlag); v. 34.
QA278.75.C75 1985 519.5 85-25044

Printed in Germany

Printing and binding: Beltz Offsetdruck, Hemsbach/Bergstr.
2147/3140-543210

To my parents

PREFACE

A full ranking of n items is simply an ordering of all these items, of the form: first choice, second choice,..., n-th choice. If two judges each rank the same n items, statisticians have used various metrics to measure the closeness of the two rankings, including Kendall's tau, Spearman's rho, Spearman's footrule, Ulam's metric, Hamming distance, and Cayley distance. These metrics have been employed in many contexts, in many applied statistical and scientific problems.

This monograph presents general methods for extending these metrics to <u>partially</u> ranked data. Here "partially ranked data" refers, for instance, to the situation in which there are n distinct items, but each judge specifies only his first through k-th choices, where $k < n$. More complex types of partially ranked data are also investigated.

Group theory is an important tool for extending the metrics. Full rankings are identified with elements of the permutation group, whereas partial rankings are identified with points in a coset space of the permutation group. The problem thus becomes one of extending metrics on the permutation group to metrics on a coset space of the permutation group. To carry out the extensions, two novel methods — the so-called Hausdorff and fixed vector methods — are introduced and implemented, which exploit this group-theoretic structure.

Various data-analytic applications of metrics on fully ranked data have been presented in the statistical literature. These can be extended now to applications of metrics on partially ranked data, and are illustrated by analyses of real data sets. The applications include fitting probability models to partially ranked data, multidimensional scaling for partially ranked data, and testing for significant differences between two populations of rankers.

* * *

It is a pleasure to acknowledge the special contribution to this monograph of a wonderful scholar and friend, Persi Diaconis. Persi has been an invaluable source of ideas and encouragement. I thank him for many hours of stimulating conversation, on the uses of mathematics in applied statistical problems.

David Pickard read the entire manuscript with great care and insight, and provided many pages of useful suggestions. I have enjoyed several enlightening discussions with him, and am sincerely grateful for all of his help.

I am indebted to Peter Huber for suggesting several intriguing areas of investigation. I thank both him and Donald Anderson for their thoughtful reading of the manuscript, and for helpful comments for improving the presentation.

Cheryl Waller did a superb job of typing the manuscript, and handled my innumerable revisions with amazing efficiency and cheerfulness.

Financial support from National Science Foundation Grant MCS80-24649 is acknowledged gratefully.

Finally, I want to thank the statistics departments at Harvard University, Stanford University, and Purdue University, for providing very substantial encouragement and assistance, during the various stages of evolution of this monograph.

TABLE OF CONTENTS

CHAPTER I - INTRODUCTION AND OUTLINE

There are many instances of partially ranked data, where several items are ranked according to some criterion, but the ordering is not complete. In its simplest form, such data arises when there are n distinct items, and each judge lists in order his k favorite items, where $k < n$. An example with $n = 5$ and $k = 3$ is afforded by the Detroit Area study, which asks people to specify the first, second, and third most important out of five named parts of marriage [B4]. More complex types of partially ranked data are also possible. The General Social Survey [D1] lists thirteen qualities that a child could possess, and from this list, respondents are asked to choose the most desirable quality, the two next most desirable qualities, the least desirable quality, and the two next least desirable qualities.

A number of interesting statistical questions arise from looking at such data. How does one measure the degree of association between two judges' partial rankings, and is the association "statistically significant"? What is a reasonable probability model for all of the respondents' rankings, and how can one test whether it fits the data? Does the data point to a significant difference between two distinct subpopulations of rankers?

This monograph approaches such questions by a novel method, which uses metrics on the permutation group and on coset spaces of the permutation group. It begins by studying, in Chapter II, some procedures which have already been developed for the case of fully ranked data. In particular, suppose that two individuals each provide a full ranking of the same set of items (i.e., a first choice, a second choice,..., a last choice). Statisticians have several ways to measure the closeness of two such rankings, including Kendall's tau, Spearman's rho, Spearman's footrule, Hamming distance, Ulam's distance, and Cayley's distance. In Chapter II, these "measures of association" for fully ranked data are identified, mathematically, with metrics on the permutation group S_n, and their invariance properties are investigated. This is a review of some material in Diaconis' important monograph [D2].

We would like to extend these "measures of association" to partially ranked data. There are various tricks which can be used to extend some of the metrics, and these are discussed in Sections III.C, IV.C, and IV.E; but the focus of this monograph is on a general procedure which enables us to extend <u>all</u> of the metrics.

In brief, the method works as follows. The set of all partial rankings of k out of n items can be identified with a quotient space S_n/S_{n-k} of the permutation group, consisting of all right cosets of the subgroup $S_{n-k} = \{\pi \varepsilon S_n: \ \pi(i) = i \ \forall i = 1,\ldots,k\}$. The problem then becomes one of extending metrics on S_n to metrics on the coset space S_n/S_{n-k} of partial rankings. This is accomplished by calculating the so-called "induced Hausdorff metrics" on S_n/S_{n-k}.

Chapter III presents the results of these calculations: each of the six aforementioned "measures of association" for fully ranked data has a natural extension to partially ranked data. Later chapters explore the appropriate generalizations to more complex types of partially ranked data, and the distributional properties of the metrics.

An important part of the monograph, Chapter VI, is concerned with data analytic applications of the metrics. Several uses of metrics on fully ranked data have already been proposed by various authors, and are presented nicely in Diaconis [D2]. These can be extended now to uses of metrics on partially ranked data, and are illustrated by analyses of real data sets.

The first data analytic application is to model building. In an article entitled "Non-Null Ranking Models," Colin Mallows [M1] proposed a useful probability model for fully ranked data. The model states that the probability of observing any permutation π decreases exponentially according to the distance from π to some fixed permutation π_0, where the distance is measured by one of the aforementioned metrics on the permutation group S_n. Mallows' model can now be extended to partially ranked data, using any of the derived metrics on the coset space S_n/S_{n-k} of partial rankings. In Chapter VI, the extension of Mallows' model is applied to some partially ranked word association data gathered by the Educational Testing Service, and it is seen to provide a good fit to the data.

Another use of the metrics is in testing whether two subpopulations of partial rankings are similar enough to be regarded as coming from a single big population of rankings. For data taking values in an arbitrary metric space, Jerome Friedman and Lawrence Rafsky [F3,F4] suggested a novel approach to this two-sample problem, based on forming the minimal spanning tree of the data. Their methodology can now be extended to partially ranked data, because any of the derived metrics on S_n/S_{n-k} can be used to construct the minimal spanning tree. The approach is applied here to some cracker preference data, to test for significant differences between the partial rankings generated by a group of preschool boys and those provided by some mothers of preschool children.

Several other applications of the metrics are discussed and illustrated in the data analysis chapter. These include multidimensional scaling for partially ranked data, and clustering partially ranked data for a goodness-of-fit test.

CHAPTER II - METRICS ON FULLY RANKED DATA

This chapter reviews, in part, an approach to permutation data presented in Diaconis' monograph, <u>Group Theory in Statistics</u> [D2].

A. Permutations: Some Important Conventions

A full ranking of n items is simply an ordering of all these items, of the form: first choice, second choice,..., n-th choice. Any such ranking can be viewed as an element π of the permutation group S_n. Here S_n is the set of all one-to-one functions from $\{1,\ldots,n\}$ onto itself, and the group operation is composition of functions.

To explicitly set up the correspondence between rankings and permutations, the items to be ranked are <u>arbitrarily</u> assigned numerical labels: item 1, item 2,..., item n. Then, for any ranking of these items, the corresponding permutation $\pi \in S_n$ is a function from $\{1,\ldots,n\}$ onto itself, whose arguments are the items, and whose values are the ranks. In other words, we have the following

<u>Convention</u>: $\pi(i)$ is the rank given to item i.
$\pi^{-1}(i)$ is the item assigned the rank i.

Finally, for any permutation π, it will be convenient to introduce the bracket notation

$$\pi = \langle\pi^{-1}(1),\ldots, \pi^{-1}(n)\rangle.$$

Since $\pi^{-1}(i)$ is the item assigned the rank i, the bracket notation corresponds to simply listing the various items in their ranked order.

A good general reference on permutations, and the permutation group S_n, is Herstein's book [H2].

B. Metrics on Permutations: Discussion and Examples

Now suppose there are two judges, who each rank the same n items. For various statistical purposes (discussed in Chapter VI), we may want to measure how close their rankings are. To do this, let $d:S_n \times S_n \to \mathbb{R}$ be a metric on the permutation group S_n, so d satisfies the usual axioms:

$$d(\pi,\pi) = 0 \qquad \forall \pi \varepsilon S_n$$

$$d(\pi,\sigma) > 0 \qquad \forall \pi,\sigma \varepsilon S_n \text{ such that } \pi \neq \sigma$$

$$d(\pi,\sigma) = d(\sigma,\pi) \quad \forall \pi,\sigma \varepsilon S_n$$

$$d(\pi,\sigma) \leq d(\pi,\tau) + d(\tau,\sigma) \qquad \forall \pi,\sigma,\tau \varepsilon S_n.$$

Let $\pi,\sigma \varepsilon S_n$ be the permutations corresponding to the two judges' rankings. Then $d(\pi,\sigma)$ can be thought of as a measure of the closeness of the two rankings, and is often called a nonparametric "measure of association."

By suitable choice of a metric on S_n, we obtain some well-known "measures of association" of two permutations, which have been used in various statistical problems:

$$R(\pi,\sigma) = \left(\sum_{i=1}^{n} (\pi(i) - \sigma(i))^2 \right)^{\frac{1}{2}}$$ is Spearman's rho (i.e., Spearman's rank correlation);

$$F(\pi,\sigma) = \sum_{i=1}^{n} |\pi(i) - \sigma(i)| \text{ is Spearman's footrule;}$$

$T(\pi,\sigma)$ = number of pairs of items, (i,j), such that $\pi(i) < \pi(j)$ and $\sigma(i) > \sigma(j)$

= minimum number of pairwise adjacent transpositions needed to transform π into σ

is Kendall's tau;

$H(\pi,\sigma) = \#\{i=1,\ldots,n: \ \pi(i) \neq \sigma(i)\}$ is Hamming distance;

$U(\pi,\sigma)$ = n - length of the longest increasing subsequence in $\sigma\pi^{-1}(1),\ldots, \sigma\pi^{-1}(n)$

= n - the maximal number of items ranked in the same order by π and σ

is Ulam's distance;

$C(\pi,\sigma)$ = n - number of cycles in $\sigma\pi^{-1}$

= minimum number of transpositions needed to transform π into σ

is Cayley's distance.

In the preceding definitions of the metrics, a few points require clarification:

(1) For any finite set Ω, $\#\Omega$ denotes the number of elements in Ω. Thus, Hamming distance is simply the number of items which are assigned different ranks by the two judges.

(2) In the alternative definitions for Kendall's tau and Cayley's distance, the aforementioned "transpositions" are to be performed

on the bracket-notation versions of π and σ, discussed in Section A. For example, if $\pi,\sigma\varepsilon S_5$ are the permutations

$$\pi = \langle 2,3,1,5,4\rangle$$

and

$$\sigma = \langle 1,3,2,5,4\rangle,$$

then $C(\pi,\sigma) = 1$ (since transposing items 1 and 2 above transforms π into σ), whereas $T(\pi,\sigma) = 3$ (since we need three pairwise adjacent transpositions to transform π into σ).

(3) In the definition of Cayley's distance, a cycle of the permutation $\sigma\pi^{-1}$ is any set of numbers $(e_1,\ldots,e_q)$ such that

$$\sigma\pi^{-1}(e_1) = e_2,\ \sigma\pi^{-1}(e_2) = e_3,\ldots,\ \sigma\pi^{-1}(e_{q-1}) = e_q,\ \sigma\pi^{-1}(e_q) = e_1.$$

It is well-known that any permutation decomposes uniquely into disjoint cycles [H2], so Cayley's distance is well-defined.

(4) The more commonly used forms of Spearman's rho, Spearman's footrule, and Kendall's tau are obtained by applying an affine transformation to the above. The transformation is to replace the metric d by 1-2d/M, where $M = M(d) = \max_{\pi,\sigma\varepsilon S_n} d(\pi,\sigma)$. This has the advantage of renormalizing the possible values to lie between -1 and 1, so that they can be interpreted in roughly the same way as a correlation coefficient.

In this monograph, however, the original metric versions d will be used, rather than the transformed versions 1-2d/M. Since the two versions are affine functions of each other, they each convey the same information. Furthermore, it will be much more convenient to work with the metric versions, both for the purpose of extending these "measures of association" to partially ranked data, and for the data-analytic applications discussed in Chapter VI.

* * *

In addition to the six metrics R, F, T, H, U, and C, a raft of other metrics on S_n can be defined. While a systematic theory of such metrics is lacking, the beginnings of a classification are in Diaconis [D2].

We have chosen to work with the six metrics presented here, because they are the most widely used. Spearman's rho [S4] and Kendall's tau have been studied extensively in the statistical literature [K6], especially in Kendall's book [K3], which also considers Spearman's footrule [S5], but dismisses it because of analytical difficulties in dealing with its sampling distribution. Interest in the footrule has been revived recently due to the paper of Diaconis and Graham [D3], which derives several of its properties. The same paper discusses Cayley's distance [C2], as does Feller [F2]. Hamming distance is widely employed in coding theory to measure the discrepancy between two binary sequences [B1,C1], while Ulam's metric is used in DNA research to measure the distance between two strings of molecules [U1,U2,B2,W1]. A. D. Gordon discusses Ulam's metric from a statistician's point of view [G2], and some of its more difficult distributional properties are investigated in the paper of Logan and Shepp [L2].

C. The Requirement of Right-Invariance

In Section A, to set up the correspondence between rankings and permutations, the first step was to assign <u>arbitrary</u> numerical labels to the n items: item 1, item 2,..., item n.

It is natural to insist that our distances between rankings should not depend on this arbitrary labeling of the items. Mathematically, this amounts to requiring that our metrics be right-invariant. That is, they must satisfy the condition

$$d(\pi,\sigma) = d(\pi\tau,\sigma\tau) \qquad \forall\pi,\sigma,\tau\varepsilon S_n.$$

Here $\tau\varepsilon S_n$ is an arbitrary relabeling of the n items, and the right-invariance property says that $d(\pi,\sigma)$ does not depend on this arbitrary relabeling.

It is easily checked that each of the six metrics introduced in Section B is, indeed, right-invariant.

* * *

As a simple illustration of the necessity of right-invariance, suppose that two individuals, Paul and Sam, each rank five flavors of ice cream. The five flavors are given the <u>arbitrary</u> numerical labels:

Chocolate = Flavor #1
Coffee = Flavor #2
Pistachio = Flavor #3
Strawberry = Flavor #4
Vanilla = Flavor #5.

If Paul ranks the flavors in the order

Coffee, Pistachio, Strawberry, Chocolate, Vanilla,

and if Sam ranks the flavors in the order

Strawberry, Vanilla, Coffee, Pistachio, Chocolate,

then, in the bracket notation of Section A, the permutations corresponding to Paul's and Sam's rankings are

$$\pi = \langle 2,3,4,1,5 \rangle$$

and

$$\sigma = \langle 4,5,2,3,1 \rangle,$$

respectively.

Now suppose instead that Chocolate is labeled Flavor #2 and Coffee Flavor #1, while all of the other labels remain the same. The rankings for Paul and Sam would now be written:

$$\pi' = \langle 1,3,4,2,5 \rangle, \quad \sigma' = \langle 4,5,1,3,2 \rangle.$$

It is natural to require that $d(\pi,\sigma) = d(\pi',\sigma')$, because in each case the flavor ranked first by Paul is ranked third by Sam, the flavor ranked second by Paul is ranked fourth by Sam, and so on -- the only change is a relabeling of the flavors. In this particular example, $\pi' = \pi\tau$ and $\sigma' = \sigma\tau$, where $\tau \varepsilon S_n$ is the permutation which exchanges 1 and 2 and leaves everything else fixed. Similarly, insisting that the distance should be invariant under <u>all</u> such arbitrary relabelings of the flavors amounts to requiring that

$$d(\pi,\sigma) = d(\pi\tau,\sigma\tau)$$

should hold for all $\pi,\sigma,\tau \varepsilon S_n$.

CHAPTER III - METRICS ON PARTIALLY RANKED DATA: THE CASE WHERE EACH JUDGE LISTS HIS k FAVORITE ITEMS OUT OF n

An illustration of this type of partially ranked data, with $n = 5$ and $k = 3$, is obtained by modifying the example of Section II.C: there are still five flavors of ice cream, but now each judge specifies only his first, second, and third choices. In this chapter, the six metrics of Section II.B are extended to metrics on this type of partially ranked data. Furthermore, the extensions preserve the necessary property of right-invariance, discussed in Section II.C.

Two methods are presented for carrying out the extensions: the so-called Hausdorff and fixed vector approaches. These methods utilize the coset space representation of partially ranked data developed in Section A. They will be related to a third method, the tied ranks approach, in Section IV.E.

The metrics introduced here will be useful for the data analytic purposes discussed in Chapters I and VI.

A. The Coset Space S_n/S_{n-k}

Let S_{n-k} be the subgroup of S_n consisting of all permutations which leave the first k integers fixed:

$$S_{n-k} = \{\pi \varepsilon S_n\colon \ \pi(i) = i \text{ for all } i = 1,\dots,k\}.$$

Define an equivalence relation on S_n as follows:

two permutations $\pi,\sigma\varepsilon S_n$ are "equivalent"

$\Longleftrightarrow \pi\sigma^{-1}\varepsilon S_{n-k}$

$\Longleftrightarrow$ there exists $\tau\varepsilon S_{n-k}$ such that $\pi = \tau\sigma$.

The preceding equivalence relation partitions S_n into so-called "equivalence classes." For any $\pi\varepsilon S_n$, the equivalence class containing π consists of all permutations equivalent to it, and is thus $\{\tau\pi: \tau\varepsilon S_{n-k}\}$. This equivalence class is a subset of the permutation group, which we denote by $S_{n-k}\pi$, and call a <u>right</u> <u>coset</u> of S_{n-k}.

The coset concept is discussed extensively in most introductory books on higher algebra; see, for example, Herstein [H2].

The relevance of the coset concept to partially ranked data arises from the fact that, to each partial ranking of k out of n items, there corresponds a unique right coset of S_{n-k}, and vice versa. In other words, there is a one-to-one correspondence:

$$\text{partial rankings} \longleftrightarrow \text{right cosets.}$$

Indeed, this is the principal upshot of the following

<u>Observation</u>: Two <u>full</u> permutations $\pi,\sigma\varepsilon S_n$ belong to the same right coset of S_{n-k}

$\Longleftrightarrow \pi^{-1}(i) = \sigma^{-1}(i) \quad \forall\ i=1,\ldots,k$

$\Longleftrightarrow \pi$ and σ induce the same <u>partial</u> ranking of k out of n items.

<u>Proof</u>: To begin with, we should state what is meant by the partial ranking induced by a full ranking: if we simply ignore the fact that,

in the full ranking, an ordering has been made of the n-k least preferred items, we obtain a partial ranking of the desired form. The second equivalence above is now immediate, if we recall that $\pi^{-1}(i)$ is the item which has been given the rank i.

To prove the first equivalence, suppose that $\pi,\sigma\varepsilon S_n$ belong to the same right coset of S_{n-k}. Then there exists $\tau\varepsilon S_{n-k}$ such that $\pi = \tau\sigma$, so for all i = 1,...,k, $\pi^{-1}(i) = \sigma^{-1}(\tau^{-1}(i)) = \sigma^{-1}(i)$. Conversely, suppose $\pi^{-1}(i) = \sigma^{-1}(i)$ for all i = 1,...,k. Then $\pi(\sigma^{-1}(i)) = i$ for all i = 1,...,k, so $\pi\sigma^{-1} \varepsilon S_{n-k}$. Thus $\pi = \tau\sigma$ where $\tau = \pi\sigma^{-1} \varepsilon S_{n-k}$, so π and σ belong to the same right coset of S_{n-k}. □

Hence, each partial ranking of k out of n items can be identified with the set of all full permutations which induce it. Moreover, this set is a right coset of S_{n-k}, namely $S_{n-k}\pi$, where π is any full ranking which induces the given partial ranking.

The set of <u>all</u> such partial rankings can therefore be identified with the set of <u>all</u> such right cosets. This latter set is called the <u>coset space</u>, and is denoted S_n/S_{n-k}.

Thus, to extend a metric on fully ranked data to a metric on partially ranked data, we must extend a metric on S_n to a metric on S_n/S_{n-k}.

B. The Hausdorff Metrics on S_n/S_{n-k}

One way of extending metrics on S_n to S_n/S_{n-k} is to construct the so-called "induced Hausdorff metrics." Indeed, this is a general procedure for extending a metric on an arbitrary finite group to a metric on its coset space. Moreover, the extension preserves the necessary property of right-invariance, discussed in Section II.C.

Let G be an arbitrary finite group, let K be any subgroup of G, and let d be a right-invariant metric on G. Then d induces a right-invariant metric d^* on the coset space G/K. d^* is called the Hausdorff metric induced by d, and it measures the distance between any two right cosets $K\pi$, $K\sigma \varepsilon G/K$ by the formula

$$d^*(K\pi, K\sigma) = \max \{ \max_{\beta \varepsilon K\sigma} \min_{\alpha \varepsilon K\pi} d(\alpha,\beta), \max_{\alpha \varepsilon K\pi} \min_{\beta \varepsilon K\sigma} d(\alpha,\beta) \}.$$

To make this definition more transparent, notice that for any fixed permutation $\beta \varepsilon K\sigma$, $\min_{\alpha \varepsilon K\pi} d(\alpha,\beta)$ is just the ordinary distance from β to the set $K\pi$. Hence, $\max_{\beta \varepsilon K\sigma} \min_{\alpha \varepsilon K\pi} d(\alpha,\beta)$ is just the ordinary distance from the set $K\pi$ to the "furthest away permutation" in $K\sigma$. Similarly, $\max_{\alpha \varepsilon K\pi} \min_{\beta \varepsilon K\sigma} d(\alpha,\beta)$ is just the ordinary distance from the set $K\sigma$ to the "furthest away permutation" in $K\pi$.

For an even more intuitive description, suppose that the boxes in Figure 1 represent the partitioning of the group G into its right cosets, and that the first box is the coset $K\pi$, while the second box is the coset $K\sigma$. Roughly speaking, the Hausdorff distance between

---

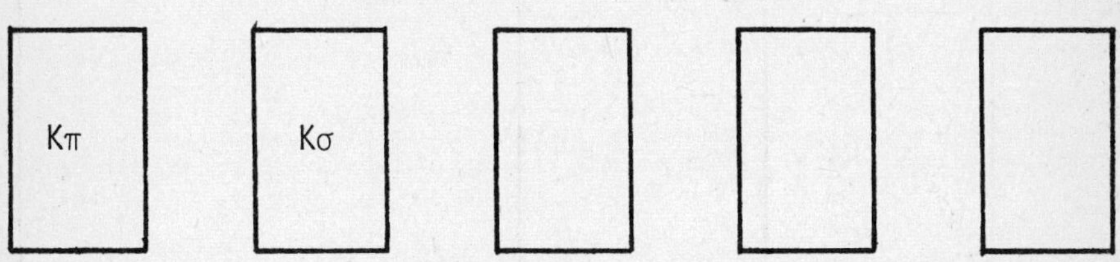

Figure 1. Partitioning of G into its Right Cosets.

$K\pi$ and $K\sigma$ is the smallest amount by which the first box must be expanded, until it includes everything in the second box, and, conversely, by which the second box must be expanded, until it includes all of the first box.

The Hausdorff metric was introduced originally in Felix Hausdorff's classic work <u>Mengenlehre</u> [H1], in the slightly different context of metrizing the entire collection of closed subsets of an arbitrary bounded metric space. It is still the simplest and most commonly used way of defining a metric on such subsets. (Other obvious candidates, such as the minimum pointwise distance $\min_{\substack{\alpha\varepsilon\Omega_1 \\ \beta\varepsilon\Omega_2}} d(\alpha,\beta)$ between the two subsets, do not, in general, satisfy the triangle inequality.) The Hausdorff metric has been studied extensively by various authors [N1,K7,K8,M3,K1].

For a treatment of the Hausdorff metric in the context of metrizing a coset space G/K, see Dieudonné [D4, Volume 2, pp. 53-54].

* * *

Taking $G = S_n$ and $K = S_{n-k}$, the Hausdorff construction enables us to extend the six metrics introduced in Section II.B to the coset space S_n/S_{n-k} of partial rankings. The following theorem, which is the main result of this chapter, gives an explicit formula for all of the induced metrics.

To state the result, it is useful to introduce the following

<u>Notation</u>: Let A, B, D, E be the following partition of the n items:

$$A = \{i = 1,\ldots,n: \ \pi(i) \le k, \ \sigma(i) \le k\}$$

$$B = \{i = 1,\ldots,n: \ \pi(i) \le k, \ \sigma(i) > k\}$$

$$D = \{i = 1,\ldots,n: \ \pi(i) > k, \ \sigma(i) \le k\}$$

$$E = \{i = 1,\ldots,n: \ \pi(i) > k, \ \sigma(i) > k\},$$

so A consists of all items ranked in the top k by both π and σ, B consists of all items ranked in the top k by π but not by σ, and so on.

Let $h = \#B = \#D$, where, as before, $\#B$ denotes the number of elements in the set B.

<u>Theorem</u>: The Hausdorff metrics on S_n/S_{n-k}, induced by the six metrics H, C, F, R, T, U of Section II.B, are:

$$H^*(S_{n-k}\pi, S_{n-k}\sigma) = \#\{i = 1,\ldots,k: \pi^{-1}(i) \ne \sigma^{-1}(i)\} + h$$

$$C^*(S_{n-k}\pi, S_{n-k}\sigma) = k - \binom{\text{number of cycles in } \sigma\pi^{-1}}{\text{consisting entirely of numbers} \le k}$$

$$F^*(S_{n-k}\pi, S_{n-k}\sigma) = \sum_{i\varepsilon A} |\pi(i) - \sigma(i)| + \sum_{i\varepsilon B} \left(\frac{n+k+1}{2} - \pi(i)\right)$$

$$+ \sum_{i\varepsilon D} \left(\frac{n+k+1}{2} - \sigma(i)\right) + h(n-k-h)$$

$$= h(2n+1-h) + \sum_{i\varepsilon A} |\pi(i) - \sigma(i)| - \sum_{i\varepsilon B} \pi(i) - \sum_{i\varepsilon D} \sigma(i)$$

$$R^*(S_{n-k}\pi, S_{n-k}\sigma) = [\sum_{i\varepsilon A} (\pi(i) - \sigma(i))^2 + h^2(n-k-h)$$

$$+\max\{ \sum_{j=1}^{h} (n+1-j-p_j)^2 + \sum_{j=1}^{h} (k+j-s_j)^2,$$

$$\sum_{j=1}^{h} (k+j-p_j)^2 + \sum_{j=1}^{h} (n+1-j-s_j)^2\}]^{1/2}$$

where $p_1 < p_2 < \ldots < p_h$ is an ordering of the set $\bigcup_{i \in B} \{\pi(i)\}$

and $s_1 < s_2 < \ldots < s_h$ is an ordering of the set $\bigcup_{i \in D} \{\sigma(i)\}$

$$T^* (S_{n-k}\pi, S_{n-k}\sigma) = \binom{\text{number of pairs of items } (i,j) \in A \times A}{\text{such that } \pi(i) < \pi(j) \text{ and } \sigma(i) > \sigma(j)}$$

$$+ h(n+k - (h-1)/2) - \sum_{i \in B} \pi(i) - \sum_{i \in D} \sigma(i)$$

$$U^* (S_{n-k}\pi, S_{n-k}\sigma) = \max \{ U(\alpha,id), U(\beta,id) \}$$

$$= n - \min \left\{ \begin{array}{l} \text{length of longest increasing subse-} \\ \text{quence in } \alpha(1),\ldots, \alpha(n); \\ \text{length of longest increasing subse-} \\ \text{quence in } \beta(1),\ldots, \beta(n) \end{array} \right\}$$

where the permutations α, β are constructed as follows:

(1) (Forced by Coset Membership)

If $i \le k$ and $\pi\sigma^{-1}(i) \le k$, let

$$\alpha^{-1}(i) = \pi\sigma^{-1}(i)$$

$$\beta(i) = \pi\sigma^{-1}(i).$$

(2) Set $\alpha(j_i) = n+1-i \quad \forall i=1,\ldots, h$

where $j_1 < \ldots < j_h$ is an ordering of

$$\{1,\ldots,k\} \cap \{\pi\sigma^{-1}(i): \; i > k\}$$

(i.e., an ordering of "the rest of $\{1,\ldots,k\}$").

Set $\beta(j_i') = n+1-i \quad \forall\, i = 1,\ldots,h$

where $j_1' < \ldots < j_h'$ is an ordering of

$$\{i \le k: \ \pi\sigma^{-1}(i) > k\}$$

(i.e., an ordering of "the rest of $\{1,\ldots,k\}$").

(3) Fill out the remaining slots of α, β with the remaining numbers in ascending order,

i.e., set $\alpha(k+i) = a_i \qquad \forall\ i = 1,\ldots,n-k$

$\beta(k+i) = b_i \qquad \forall\ i = 1,\ldots,n-k$

where $a_1 < \ldots < a_{n-k}$ is an ordering of

$$\{1,\ldots,n-h\} \setminus \{i \le k: \ \pi\sigma^{-1}(i) \le k\}$$

$b_1 < \ldots < b_{n-k}$ is an ordering of

$$\{1,\ldots,n-h\} \setminus \{\pi\sigma^{-1}(i): \ i \le k \text{ and } \pi\sigma^{-1}(i) \le k\}.$$

Remarks: $h = \#B$ is the number of items ranked in the top k by π but not by σ, so h is a measure of how much the two partial rankings do not overlap. The fact that $\#B = \#D$ is immediate from $\#A + \#B = k = \#A + \#D$.

The Hausdorff metric H^* is the sum of two pieces, each of which has a natural interpretation. The first piece,

$$\#\{i = 1,\ldots,k: \ \pi^{-1}(i) \ne \sigma^{-1}(i)\},$$

is itself a metric on S_n/S_{n-k}. It counts the number of ranks between 1 and k which have been assigned different items by π and σ. The second piece, h, can be regarded as a "penalty term" measuring the non-overlap of the two partial rankings.

Similarly, the Hausdorff metric F^* breaks up into the sum of two pieces. We can think of π and σ as assigning the "pseudo-rank" $\frac{n+k+1}{2}$ to all of the items not in their respective "top k," where $\frac{n+k+1}{2}$ is the average of the unused ranks k+1,...,n. Then the first piece of F^*,

$$\sum_{i\varepsilon A} |\pi(i) - \sigma(i)| + \sum_{i\varepsilon B} (\frac{n+k+1}{2} - \pi(i)) + \sum_{i\varepsilon D} (\frac{n+k+1}{2} - \sigma(i)),$$

is just the sum of the absolute differences in the "pseudo-ranks," and is thus a natural extension of Spearman's footrule (see also Section C). The second piece, h(n-k-h), can again be thought of as a "penalty term."

Not every Hausdorff metric has such a nice interpretation (e.g., U^*). Section IV.E will present a further discussion of these matters.

<u>Proof of Theorem:</u> The proofs for F^*, R^*, T^* and U^* are deferred to Section IV.D, where a more general theorem will be proved. The results for H^* and C^* will also follow from the theorem in IV.D; however, they are proved here as well, because for these two metrics, the proof is considerably simpler in the present case than in the case of IV.D. Furthermore, the proofs for H^* and C^* provide a nice demonstration of the power of the group-theoretic approach.

The first step in the proofs is the observation that in addition to being right-invariant, H and C are <u>left-invariant</u>, i.e.,

$$d(\pi,\sigma) = d(\tau\pi,\tau\sigma) \qquad \forall \pi,\sigma,\tau\varepsilon S_n,$$

where d = H or C. Indeed,

$$H(\tau\pi,\tau\sigma) = \#\{i = 1,\ldots,n:\ \tau\pi(i) \neq \tau\sigma(i)\}$$

$$= \#\{i = 1,\ldots,n:\ \tau^{-1}\tau\pi(i) \neq \tau^{-1}\tau\sigma(i)\}$$

$$= H(\pi,\sigma).$$

To see that Cayley's distance is left-invariant, notice that if $(e_1,\ldots,e_q)$ is a cycle of $\sigma\pi^{-1}$, then $(\tau(e_1),\ldots, \tau(e_q))$ is a cycle of $\tau\sigma\pi^{-1}\tau^{-1}$. Therefore, $\tau\sigma\pi^{-1}\tau^{-1}$ has the same number of cycles as $\sigma\pi^{-1}$, and

$$C(\tau\pi,\tau\sigma) = n - (\#\text{ cycles in } \tau\sigma\pi^{-1}\tau^{-1})$$

$$= n - (\#\text{ cycles in } \sigma\pi^{-1})$$

$$= C(\pi,\sigma).$$

The significance of the left-invariance property lies in the following lemma, due to Diaconis and Graham. It says that when d is left-invariant, the Hausdorff distance d^* between two cosets is the same as the interpoint distance between the two closest members of the cosets.

Lemma: Let G be any finite group, K any subgroup of G, d a metric on G, and d^* the induced Hausdorff metric on G/K:

$$d^*(K\pi,K\sigma) = \max\{\max_{\beta\varepsilon K\sigma}\min_{\alpha\varepsilon K\pi} d(\alpha,\beta),\ \max_{\alpha\varepsilon K\pi}\min_{\beta\varepsilon K\sigma} d(\alpha,\beta)\}$$

for all $K\pi$, $K\sigma \in G/K$. If d is left-invariant, then d^* may be computed according to the simpler formula:

$$d^*(K\pi,K\sigma) = \min_{\substack{\alpha\varepsilon K\pi \\ \beta\varepsilon K\sigma}} d(\alpha,\beta) = \min_{\beta\varepsilon K\sigma} d(\pi,\beta).$$

Remark: The lemma is not true if d is not left-invariant. As a counterexample, let $G = S_5$, let $K \subset S_5$ be the subgroup $K = S_2 = \{\gamma\varepsilon S_5: \ \gamma(i) = i \quad \forall\ i = 1,2,3\}$, and let d = T be Kendall's tau on S_5. Note $K = \{id,\tau\} \subset S_5$, where id denotes the identity permutation, and where τ is the transposition which switches 4 and 5 and leaves 1, 2 and 3 fixed.

In the bracket-notation of Section II.A, let

$$\pi = <1,2,3,4,5> \varepsilon\ S_5$$

be the identity permutation, and let

$$\sigma = <1,2,4,3,5> \varepsilon\ S_5.$$

Then the right coset $S_2\pi$ corresponds to the partial ranking which places items 1, 2, and 3 in the top three positions, while the right coset $S_2\sigma$ corresponds to the partial ranking which places items 1, 2, and 4 in the top three positions.

The interpoint distance between the two closest members of the cosets is

$$\min_{\substack{\alpha\varepsilon S_2\pi \\ \beta\varepsilon S_2\sigma}} T(\alpha,\beta) = T(\pi,\sigma) = 1.$$

But the Hausdorff distance

$$T^*(S_2\pi,S_2\sigma) = \max\{\max_{\beta\varepsilon S_2\sigma}\min_{\alpha\varepsilon S_2\pi} T(\alpha,\beta),\ \max_{\alpha\varepsilon S_2\pi}\min_{\beta\varepsilon S_2\sigma} T(\alpha,\beta)\}$$

$= \max \{ T(<1,2,3,4,5>, <1,2,4,5,3>),$

$T(<1,2,3,5,4>, <1,2,4,3,5>) \}$

$= \max \{2,2\}$

$= 2.$

Proof of Lemma: Note

$$\max_{\beta\varepsilon K\sigma} \min_{\alpha\varepsilon K\pi} d(\alpha,\beta) = \max_{k_2\varepsilon K} \min_{k_1\varepsilon K} d(k_1\pi, k_2\sigma)$$

$$= \max_{k_2\varepsilon K} (\min_{k_1\varepsilon K} d(\pi, k_1^{-1} k_2 \sigma))$$

$$= \max_{k_2\varepsilon K} (\min_{k\varepsilon K} d(\pi,k\sigma))$$

$$= \min_{k\varepsilon K} d(\pi,k\sigma)$$

$$= \min_{\beta\varepsilon K\sigma} d(\pi,\beta).$$

Similarly,

$$\max_{\alpha\varepsilon K\pi} \min_{\beta\varepsilon K\sigma} d(\alpha,\beta) = \min_{\beta\varepsilon K\sigma} d(\pi,\beta).$$

Hence $d^*(K\pi, K\sigma) = \min_{\beta\varepsilon K\sigma} d(\pi,\beta)$.

To show that $\min_{\beta\varepsilon K\sigma} d(\pi,\beta) = \min_{\substack{\alpha\varepsilon K\pi \\ \beta\varepsilon K\sigma}} d(\alpha,\beta)$, note

$$\min_{\beta\varepsilon K\sigma} d(\pi,\beta) = \min_{k\varepsilon K} d(\pi,k\sigma)$$

$$= \min_{k_1,k_2\varepsilon K} d(\pi, k_1^{-1}k_2\sigma)$$

$$= \min_{k_1,k_2 \in K} d(k_1\pi, k_2\sigma)$$

$$= \min_{\substack{\alpha\in K\pi \\ \beta\in K\sigma}} d(\alpha,\beta). \square$$

<u>Proof of Theorem for H^*</u>:

$$H^*(S_{n-k}\pi, S_{n-k}\sigma) = \min_{\beta\in S_{n-k}\sigma} H(\pi,\beta)$$

$$= \min_{\beta\in S_{n-k}\sigma} \#\{i = 1,\ldots,n: \pi(i) \neq \beta(i)\}$$

$$= \min_{\beta\in S_{n-k}\sigma} \#\{i = 1,\ldots,n: \pi^{-1}(i) \neq \beta^{-1}(i)\}$$

$$= \min_{\beta\in S_{n-k}\sigma} [\#\{i = 1,\ldots,k: \pi^{-1}(i) \neq \beta^{-1}(i)\}$$

$$+ \#\{i = k+1,\ldots,n: \pi^{-1}(i) \neq \beta^{-1}(i)\}]$$

$$= \#\{i = 1,\ldots,k: \pi^{-1}(i) \neq \sigma^{-1}(i)\}$$

$$+ \min_{\beta\in S_{n-k}\sigma} \#\{i = k+1,\ldots,n: \pi^{-1}(i) \neq \beta^{-1}(i)\}.$$

The first equality is by the lemma, and the last follows from the fact that $\sigma^{-1}(i) = \beta^{-1}(i)$ for all $i \leq k$ by the Observation in Section A. It remains to show that $h = m$, where

$$m = \min_{\beta\in S_{n-k}\sigma} \#\{i = k+1,\ldots,n: \pi^{-1}(i) \neq \beta^{-1}(i)\}.$$

Let $b_1,\ldots,b_h$ and $d_1,\ldots,d_h$ be enumerations of the sets B and D, respectively. Let β_0 be the permutation given by

$$\beta_0(i) = \sigma(i) \qquad \forall i \in A \cup D$$

$$\beta_0(i) = \pi(i) \qquad \forall i \in E$$

$$\beta_0(b_j) = \pi(d_j) \qquad \forall j=1,\ldots,h.$$

Then $\beta_0 \in S_{n-k}\sigma$ and

$$\begin{aligned}\#\{i=k+1,\ldots,n: \pi^{-1}(i) \neq \beta_0^{-1}(i)\} &= \#\,\beta_0(B) \\ &= \#B \\ &= h.\end{aligned}$$

Thus $m \leq h$. On the other hand, for any $\beta \in S_{n-k}\sigma$,

$$\{\, i=k+1,\ldots,n: \pi^{-1}(i) \neq \beta^{-1}(i)\} \supset \beta(B)$$

and $\beta(B)$ has cardinality h, so $m \geq h$. □

<u>Proof of Theorem for C^*</u>: For any permutation σ, define a cycle of σ to be of type 1 if all its numbers are $\leq k$, and of type 2 if some number in the cycle is $> k$. By the right-invariance of C^*, it suffices to prove the result when π = id, i.e. to show

$$C^*(S_{n-k}, S_{n-k}\sigma) = k - (\#\text{ of type 1 cycles in }\sigma).$$

For any $\beta \in S_{n-k}\sigma$, $\sigma^{-1}(i) = \beta^{-1}(i)$ for all $i \leq k$. Thus any type 1 cycle of σ is also a type 1 cycle of β, and vice versa. Hence

$$\begin{aligned}
C^*(S_{n-k}, S_{n-k}\sigma) &= \min_{\beta \varepsilon S_{n-k}\sigma} C(id,\beta) \\
&= n - \max_{\beta \varepsilon S_{n-k}\sigma} (\#\text{ cycles in } \beta) \\
&= n - \max_{\beta \varepsilon S_{n-k}\sigma} [(\#\text{ of type 1 cycles in } \beta) \\
&\qquad + (\#\text{ of type 2 cycles in } \beta)] \\
&= n - (\#\text{ of type 1 cycles in } \sigma) \\
&\qquad - \max_{\beta \varepsilon S_{n-k}\sigma} (\#\text{ of type 2 cycles in } \beta).
\end{aligned}$$

It remains to show $m = n - k$, where $m = \max_{\beta \varepsilon S_{n-k}\sigma}$ (# of type 2 cycles in β).

Let $\beta \varepsilon S_{n-k}\sigma$. There are only n-k integers $> k$ and $\leq n$, and each occurs in only one cycle of β, so $m \leq n-k$. On the other hand, define the permutation β_0 to have the same cycle structure as σ, except that those type 2 cycles of σ with <u>more than one</u> member $> k$ are modified in the cycle representation of β_0, as follows: without loss of generality, any such cycle of σ may be written

$$(d_1, e_1, e_2, \ldots, e_{q_1}, d_2, e_{q_1+1}, \ldots, e_{q_2}, d_3, \ldots, d_r, e_{q_{r-1}+1}, \ldots, e_{q_r})$$

where $d_i > k$ and $e_i \leq k$ for all i. In β_0, the corresponding numbers d_i and e_i should lie in r disjoint cycles:

$$(d_1, e_1, e_2, \ldots, e_{q_1}),\ (d_2, e_{q_1+1}, \ldots, e_{q_2}), \ldots, (d_r, e_{q_{r-1}+1}, \ldots, e_{q_r}).$$

Note $\beta_0 \;\varepsilon\; S_{n-k}\sigma$, since $\beta_0^{-1}(i) = \sigma^{-1}(i)\ \ \forall\ i \leq k$. Moreover, any type 2 cycle of β_0 contains exactly one number $> k$. Thus β_0 has n-k type 2 cycles, so $m \geq n-k$. $\square$

C. The Fixed Vector Metrics on S_n/S_{n-k}

A second approach to metrizing certain coset spaces has been suggested by Andrew Rukhin [R1]. It is based on the following

Proposition: Let G be any group, K any subgroup of G, and let $\|\cdot\|$ denote a norm on $\mathbb{R}^n$. Let ρ be a group representation of G; that is, ρ assigns to each $\pi\varepsilon G$ a non-singular linear transformation

$$\rho(\pi): \mathbb{R}^n \to \mathbb{R}^n,$$

in such a way that

$$\rho(\pi\sigma) = \rho(\pi)\rho(\sigma) \quad \forall\pi,\sigma\varepsilon G.$$

Suppose further that ρ is "norm preserving" in the sense that

$$\|\rho(\pi)^T \underset{\sim}{w}\| = \|\underset{\sim}{w}\| \quad \forall \underset{\sim}{w}\,\varepsilon\mathbb{R}^n \quad \forall\pi\varepsilon G.$$

Finally, suppose $\underset{\sim}{v}\varepsilon\mathbb{R}^n$ is a vector fixed by K under ρ^T, but not fixed by any larger subgroup. That is, suppose

$$\rho(\pi)^T \underset{\sim}{v} = \underset{\sim}{v} \quad \forall\pi\varepsilon K$$

and

$$\rho(\pi)^T \underset{\sim}{v} \neq \underset{\sim}{v} \quad \forall\pi\varepsilon G\backslash K.$$

If the preceding conditions are satisfied, then

$$d_{fv}(K\pi,K\sigma) = \|(\rho(\pi)-\rho(\sigma))^T \underset{\sim}{v}\|$$

is a right-invariant metric on the coset space G/K.

<u>Remark</u>: d_{fv} is called the "fixed vector metric" on G/K.

<u>Proof of Proposition</u>: First of all, we must show that d_{fv} is well-defined and does not depend on the choice of π and σ as the "representative elements" of their respective cosets. This follows from the fact that $\underset{\sim}{v}$ is fixed by K under ρ^T. For if $k_1\pi$ and $k_2\sigma$ are any other members of the cosets $K\pi$ and $K\sigma$, then

$$\begin{aligned} \|(\rho(k_1\pi)-\rho(k_2\sigma))^T\underset{\sim}{v}\| &= \|\rho(k_1\pi)^T\underset{\sim}{v}-\rho(k_2\sigma)^T\underset{\sim}{v}\| \\ &= \|\rho(\pi)^T\rho(k_1)^T\underset{\sim}{v}-\rho(\sigma)^T\rho(k_2)^T\underset{\sim}{v}\| \\ &= \|\rho(\pi)^T\underset{\sim}{v}-\rho(\sigma)^T\underset{\sim}{v}\| \\ &= \|(\rho(\pi)-\rho(\sigma))^T\underset{\sim}{v}\| . \end{aligned}$$

It is also easy to check that d_{fv} satisfies the axioms for a metric (listed on page 6), because $\|\cdot\|$ satisfies the corresponding conditions for a vector norm. The least straightforward axiom to verify is positivity:

$$d(K\pi,K\sigma) > 0 \quad \forall K\pi,K\sigma\in G/K \text{ such that } K\pi \neq K\sigma.$$

This is a consequence of the assumption that $\underset{\sim}{v}$ is not fixed under ρ^T by any subgroup larger than K. Indeed, suppose $K\pi,K\sigma\in G/K$ are such that $K\pi \neq K\sigma$ but $d_{fv}(K\pi,K\sigma) = 0$. Then $\sigma = g\pi$, where $g = \sigma\pi^{-1}\in G\setminus K$, and

$$\begin{aligned} 0 &= d_{fv}(K\pi,K\sigma) \\ &= \|(\rho(\pi)-\rho(\sigma))^T\underset{\sim}{v}\| \\ &= \|(\rho(\pi)-\rho(g\pi))^T\underset{\sim}{v}\| \\ &= \|(\rho(\pi)-\rho(g)\rho(\pi))^T\underset{\sim}{v}\| \end{aligned}$$

$$= \|\rho(\pi)^T(\underset{\sim}{v}-\rho(g)^T\underset{\sim}{v})\|$$

$$= \|\underset{\sim}{v}-\rho(g)^T\underset{\sim}{v}\|,$$

whence $\underset{\sim}{v}$ is fixed by g under ρ^T, a contradiction.

Finally, d_{fv} is right-invariant because ρ is "norm preserving." Indeed, for any $\pi,\sigma,\tau\varepsilon G$ we have

$$\begin{aligned} d_{fv}(K\pi\tau,K\sigma\tau) &= \|(\rho(\pi\tau)-\rho(\sigma\tau))^T\underset{\sim}{v}\| \\ &= \|\rho(\tau)^T(\rho(\pi)-\rho(\sigma))^T\underset{\sim}{v}\| \\ &= \|(\rho(\pi)-\rho(\sigma))^T\underset{\sim}{v}\| \\ &= d_{fv}(K\pi,K\sigma). \quad \square \end{aligned}$$

Of course, for this construction to be useful, there must exist a fixed vector $\underset{\sim}{v}$. This topic is treated in Appendix A, which gives precise conditions that must be satisfied by the representation ρ, in order for $\underset{\sim}{v}$ to exist. Appendix A also helps to put the fixed vector method "in context," by introducing the elaborate mathematical machinery associated with group representations and fixed vectors.

* * *

The statistical significance of the fixed vector metric stems largely from the fact that it gives such nice results in a few simple cases. In particular, Example 2 below gives rise to several metrics that are useful for analyzing partially ranked data.

Example 1 (due to Persi Diaconis): Take $G = S_n$, take $K = \{id\}$ to be the trivial subgroup, and let $\|\cdot\|$ be the ℓ^1 norm on $\mathbb{R}^n$. Let ρ be the n-dimensional permutation representation. Thus, for each $\pi\varepsilon S_n$, $\rho(\pi)$ is the permutation matrix given by

$$\rho(\pi)_{ij} = \delta_{i\pi(j)} = \begin{cases} 1 \text{ if } i = \pi(j) \\ 0 \text{ if } i \neq \pi(j). \end{cases}$$

Note $\rho(\pi)^T$ is "norm preserving" because it merely permutes the components of any vector $\underset{\sim}{w}$:

$$(*) \qquad (\rho(\pi)^T\underset{\sim}{w})_i = \sum_{j=1}^{n} \rho(\pi)_{ji}w_j = \sum_{j=1}^{n} \delta_{j\pi(i)}w_j = w_{\pi(i)} \quad \forall\pi\varepsilon S_n \quad \forall \underset{\sim}{w}\varepsilon \mathbb{F}$$

Finally, let the fixed vector $\underset{\sim}{v} = (1,2,\ldots,n)^T$.

Then the fixed vector metric becomes Spearman's footrule:

$$\begin{aligned} d_{fv}(\pi,\sigma) &= \|(\rho(\pi)-\rho(\sigma))^T\underset{\sim}{v}\|_1 \\ &= \sum_{i=1}^{n} |(\rho(\pi)^T\underset{\sim}{v})_i-(\rho(\sigma)^T\underset{\sim}{v})_i| \\ &= \sum_{i=1}^{n} |v_{\pi(i)}-v_{\sigma(i)}| \\ &= \sum_{i=1}^{n} |\pi(i)-\sigma(i)| \\ &= F(\pi,\sigma). \end{aligned}$$

<u>Example 2</u>: If G, $\|\cdot\|$, and ρ are as in Example 1, but if we now take

$$K = S_{n-k} = \{\pi\varepsilon S_n: \pi(i) = i \quad \forall i = 1,\ldots,k\},$$

then it is natural to regard the resulting fixed vector metric as an <u>extension</u> of Spearman's footrule to the coset space S_n/S_{n-k} of partial rankings. It remains to specify the vector $\underset{\sim}{v}$ that is fixed by S_{n-k}

under ρ^T, but not fixed by any larger subgroup. Appendix A treats the general question of the existence of such fixed vectors, but in the present example, it is easy to see that there are many possible choices of $\underset{\sim}{v}$.

One especially noteworthy choice is to take

$$\underset{\sim}{v} = (1,2,\ldots,k, \frac{n+k+1}{2},\ldots, \frac{n+k+1}{2})^T.$$

The resulting metric on S_n/S_{n-k} is

$$F_{fv}(S_{n-k}\pi, S_{n-k}\sigma) = \sum_{i\varepsilon A} |\pi(i)-\sigma(i)| + \sum_{i\varepsilon B} (\frac{n+k+1}{2} - \pi(i)) + \sum_{i\varepsilon D} (\frac{n+k+1}{2} - \sigma(i))$$

This is the same as the "first piece" of the Hausdorff metric F^*, as discussed in Section B. It has the particularly nice interpretation that π and σ are assigning rank $\frac{n+k+1}{2}$ to those items not in their "top k," where $\frac{n+k+1}{2}$ is the average of the unused ranks $k+1,\ldots,n$.

More generally, by (*), we can choose the fixed vector $\underset{\sim}{v}$ to be any $\underset{\sim}{v} = (v_1,\ldots,v_n)^T$ such that $v_1,\ldots,v_{k+1}$ are distinct and $v_{k+1} = v_{k+2} = \ldots = v_n$ The resulting metric on S_n/S_{n-k} is

$$F_{fv,\underset{\sim}{v}}(S_{n-k}\pi, S_{n-k}\sigma) = \sum_{i=1}^{n} |v_{\pi(i)} - v_{\sigma(i)}|.$$

This has the interpretation that the two judges are assigning the "pseudo-ranks" $v_{\pi(i)}$ and $v_{\sigma(i)}$ to item i, and then computing Spearman's footrule using the pseudo-ranks rather than the ordinary ranks.

Finally, all of the above can be carried out with different norms. For example, if we substitute the (euclidean) ℓ^2 norm on $\mathbb{R}^n$, the fixed vector metric becomes Spearman's rho when $K = \{id\}$, and the analogous "pseudo-rank" extension of Spearman's rho when $K = S_{n-k}$:

$$R_{fv,\underset{\sim}{v}}(S_{n-k}\pi,\ S_{n-k}\sigma) = (\sum_{i=1}^{n}(v_{\pi(i)} - v_{\sigma(i)})^2)^{\frac{1}{2}}.$$

Using other ℓ^p norms, for $p \in [1,\infty]$, produces equally plausible metrics.

CHAPTER IV - METRICS ON OTHER TYPES OF PARTIALLY RANKED DATA

In this chapter, the results of Chapter III are extended to the following more general type of partially ranked data: each judge is given a list of n items. He partitions these n items into r groups: the first group contains his n_1 favorite items, the second group contains his n_2 next most preferred items, and so on; the final group contains his n_r least favorite items. The judge does not state any preferences among members of the same group. Here $n_1,\ldots,n_r$ are strictly positive integers, satisfying $\sum_{i=1}^{r} n_i = n$; the partial ranking is said to be "of type $n_1,\ldots,n_r$."

The partially ranked data considered in Chapter III is actually a special case of the kind described here, with

$$r = k + 1,\ n_1 = \ldots = n_k = 1, \text{ and } n_{k+1} = n\text{-}k.$$

Just as before, there is a coset space representation for this kind of partially ranked data, and it is discussed in Section A. Sections B and C present the Hausdorff and fixed vector metrics, respectively. These distances are calculated between two partial rankings which are both of the <u>same</u> type $n_1,\ldots,n_r$.

Section D tackles the problem of nonparametrically measuring the association between two partial rankings of <u>different</u> types. It is shown that Hausdorff distances can still be computed in this most general situation. Complete proofs are presented, which also cover many of the special cases considered in Sections III.B and IV.B.

Finally, Section E discusses the so-called "tied ranks approach" to metrizing partially ranked data, and compares it with the Hausdorff and fixed vector approaches.

A. The Coset Space S_n/S, Where $S = S_{n_1} \times S_{n_2} \times \ldots \times S_{n_r}$

Let $N_1,\ldots,N_r$ be the following partition of $\{1,\ldots,n\}$:

$$\begin{aligned} N_1 &= \{1,2,\ldots,n_1\} \\ N_2 &= \{n_1+1,\ldots,n_1+n_2\} \\ N_3 &= \{n_1+n_2+1,\ldots,n_1+n_2+n_3\} \\ &\vdots \\ N_r &= \{n_1+\ldots+n_{r-1}+1,\ldots,n\} \end{aligned}$$

and let $S_{n_1}, S_{n_2},\ldots,S_{n_r}$ be the subgroups of S_n given by:

$$\begin{aligned} S_{n_1} &= \{\pi \varepsilon S_n: \pi(i) = i \quad \forall\, i \notin N_1\} \\ S_{n_2} &= \{\pi \varepsilon S_n: \pi(i) = i \quad \forall\, i \notin N_2\} \\ &\vdots \\ S_{n_r} &= \{\pi \varepsilon S_n: \pi(i) = i \quad \forall\, i \notin N_r\}. \end{aligned}$$

Let S denote the subgroup $S_{n_1} \times S_{n_2} \times \ldots \times S_{n_r}$ of S_n. Thus S is the subgroup

$$S = \{\pi\varepsilon S_n: \pi(N_i) = N_i \quad \forall\ i = 1,\ldots,r\}$$

of all rankings which permute the first n_1 items among the first n_1 ranks, and which permute the next n_2 items among the next n_2 ranks, and so on.

Analogously to Section III.A, there is a one-to-one correspondence:

$$\begin{matrix}\text{partial rankings} \\ \text{of type } n_1,\ldots,n_r\end{matrix} \longleftrightarrow \text{right cosets of } S.$$

Indeed, this is the whole point of the following

<u>Observation</u>: Two <u>full</u> permutations $\pi,\sigma\varepsilon S_n$ belong to the same right coset of S

$$\Longleftrightarrow \pi^{-1}(N_i) = \sigma^{-1}(N_i) \text{ for all } i = 1,\ldots,r$$

$$\Longleftrightarrow \pi \text{ and } \sigma \text{ induce the same } \underline{\text{partial}} \text{ ranking of type } n_1,\ldots,n_r.$$

<u>Proof</u>: The second equivalence follows directly from the fact that $\pi^{-1}(N_i)$ is the collection of items which are given the ranks in the set N_i.

To prove the first equivalence, suppose that $\pi,\sigma\varepsilon S_n$ belong to the same right coset of S. Then there exists $\tau\varepsilon S$ such that $\pi = \tau\sigma$, so for all $i = 1,\ldots,r$,

$$\tau^{-1}(N_i) = N_i$$

and

$$\pi^{-1}(N_i) = \sigma^{-1}(\tau^{-1}(N_i)) = \sigma^{-1}(N_i).$$

Conversely, suppose

$$\pi^{-1}(N_i) = \sigma^{-1}(N_i) \text{ for all } i = 1,\ldots,r.$$

Then

$$\pi(\sigma^{-1}(N_i)) = N_i \text{ for all } i = 1,\ldots,r,$$

so $\pi\sigma^{-1} \varepsilon S$. Hence $\pi = \tau\sigma$ where $\tau = \pi\sigma^{-1} \varepsilon S$, so π and σ belong to the same right coset of S.$\square$

Thus, each partial ranking of type $n_1,\ldots,n_r$ is identified with the set of full rankings which induce it. Furthermore, this set is a right coset of S, namely $S\pi$, where π is any full ranking which induces the given partial ranking.

To extend a metric on fully ranked data to a metric on partially ranked data of type $n_1,\ldots,n_r$, we must extend a metric on S_n to a metric on the coset space S_n/S.

B. The Hausdorff Metrics on S_n/S

The Hausdorff metrics on S_n/S are defined by taking $G = S_n, K = S$ in the definition of Section III.B, page 15.

<u>Theorem</u>: Let n_{ij} be the number of elements in the set

$$\pi^{-1}(N_i) \cap \sigma^{-1}(N_j),$$

so n_{ij} is the number of items placed in the i-th category by the first judge and in the j-th category by the second. Then the Hausdorff distances on S_n/S, induced by the six metrics of Section II.B, are:

$$H^*(S\pi,S\sigma) = n - \sum_{i=1}^{r} n_{ii}$$

$$T^*(S\pi,S\sigma) = \sum_{\substack{i<i' \\ j\geq j'}} n_{ij}n_{i'j'} = \sum_{\substack{i\leq i' \\ j>j'}} n_{ij}n_{i'j'}$$

$$U^*(S\pi,S\sigma) = n - \min\{ \max_{\substack{i(j)\text{ a non-decreasing} \\ \text{function of } j}} \sum_{i=1}^{r} n_{i(j)j}, \quad \max_{\substack{j(i)\text{ a non-decreasing} \\ \text{function if } i}} \sum_{j=1}^{r} n_{ij(i)} \}$$

$$F^*(S\pi,S\sigma) = \max\{ \sum_{i=1}^{r}\sum_{j=1}^{r} n_{ij} \left| \sum_{k=1}^{i-1} n_k - \sum_{k=1}^{j-1} n_k + \sum_{k=1}^{j-1} n_{ik} - \sum_{k=i+1}^{r} n_{kj} \right|,$$

$$\sum_{i=1}^{r}\sum_{j=1}^{r} n_{ij} \left| \sum_{k=1}^{i-1} n_k - \sum_{k=1}^{j-1} n_k + \sum_{k=j+1}^{r} n_{ik} - \sum_{k=1}^{i-1} n_{kj} \right| \}$$

$$R^*(S\pi,S\sigma) = \max\{ \sum_{i=1}^{r}\sum_{j=1}^{r} n_{ij} \left(\sum_{k=1}^{i-1} n_k - \sum_{k=1}^{j-1} n_k + \sum_{k=1}^{j-1} n_{ik} - \sum_{k=i+1}^{r} n_{kj} \right)^2,$$

$$\sum_{i=1}^{r}\sum_{j=1}^{r} n_{ij} \left(\sum_{k=1}^{i-1} n_k - \sum_{k=1}^{j-1} n_k + \sum_{k=j+1}^{r} n_{ik} - \sum_{k=1}^{i-1} n_{kj} \right)^2 \}^{\frac{1}{2}}$$

$$C^*(S\pi,S\sigma) = \begin{cases} n_{12} & \text{if } r = 2 \\ n - \left(\sum_{i=1}^{3} n_{ii} + \sum_{i<j} \min\{n_{ij},n_{ji}\} + |n_{12} - n_{21}| \right) & \text{if } r = 3. \end{cases}$$

<u>Notation</u>: $\sum_{i<j}$ denotes summation over ordered pairs (i,j) such that $i < j$, and <u>not</u> $\sum_{i=1}^{j-1}$. Similarly, $\sum_{\substack{i<i' \\ j\geq j'}}$ denotes summation over ordered quadruples (i,i',j,j') such that $i < i'$ and $j \geq j'$.

Before proceeding with the proof of the theorem, it may be helpful to consider the following

Example: Let π^* be the partial ranking which places items 1, 3, and 7 in the first category, items 2, 4, and 6 in the second category, and items 5 and 8 in the third category. Thus, π^* is a partial ranking of $n = 8$ items, of type $n_1 = 3$, $n_2 = 3$, $n_3 = 2$. We denote this partial ranking by

$$\pi^* = \langle(1,3,7), (2,4,6), (5,8)\rangle,$$

where the asterisk on π^* indicates that it is a partial ranking, and where the rounded parentheses denote ties within each category.

Let σ^* be the partial ranking

$$\sigma^* = \langle(1,2,4), (3,5,6), (7,8)\rangle.$$

Then π^* and σ^* can be identified with the cosets $S\pi$ and $S\sigma$, where $S = S_3 \times S_3 \times S_2$, and where π and σ are the full rankings

$$\pi = \langle 1,3,7,2,4,6,5,8\rangle$$

and

$$\sigma = \langle 1,2,4,3,5,6,7,8\rangle,$$

respectively.

The parameters n_{ij} (as defined in the statement of the theorem) are:

$$\begin{array}{lll} n_{11} = 1 & n_{12} = 1 & n_{13} = 1 \\ n_{21} = 2 & n_{22} = 1 & n_{23} = 0 \\ n_{31} = 0 & n_{32} = 1 & n_{33} = 1 \end{array}$$

Hence, the Hausdorff distances between the two partial rankings are given by:

$$H^*(S\pi,S\sigma) = n - \sum_{i=1}^{3} n_{ii} = 8 - (1+1+1) = 5$$

$$T^*(S\pi,S\sigma) = \sum_{\substack{i<i' \\ j\geq j'}} n_{ij}\, n_{i'j'}$$

$$= n_{11}(n_{21}+n_{31}) + n_{12}(n_{21}+n_{22}+n_{31}+n_{32})$$

$$+ n_{13}(n_{21}+n_{22}+n_{23}+n_{31}+n_{32}+n_{33})$$

$$+ n_{21}n_{31} + n_{22}(n_{31}+n_{32}) + n_{23}(n_{31}+n_{32}+n_{33})$$

$$= 1\cdot 2 + 1\cdot 4 + 1\cdot 5 + 2\cdot 0 + 1\cdot 1 + 0\cdot 2 = 12$$

$$U^*(S\pi,S\sigma) = n - \min\{n_{21}+n_{32}+n_{33},\ n_{11}+n_{21}+n_{32}\}$$

$$= 8 - \min\{4,4\} = 4$$

$$F^*(S\pi,S\sigma) = \max\{1\cdot 2 + 1\cdot 4 + 1\cdot 5 + 2\cdot 3 + 1\cdot 1 + 0 + 0 + 1\cdot 3 + 1\cdot 1,$$

$$1\cdot 2 + 1\cdot 2 + 1\cdot 6 + 2\cdot 3 + 1\cdot 1 + 0 + 0 + 1\cdot 2 + 1\cdot 1\}$$

$$= \max\{22,20\} = 22$$

$$R^*(S\pi,S\sigma) = \max\{1\cdot 2^2 + 1\cdot 4^2 + 1\cdot 5^2 + 2\cdot 3^2 + 1\cdot 1^2 + 0 + 0 + 1\cdot 3^2 + 1\cdot 1^2,$$

$$1\cdot 2^2 + 1\cdot 2^2 + 1\cdot 6^2 + 2\cdot 3^2 + 1\cdot 1^2 + 0 + 0 + 1\cdot 2^2 + 1\cdot 1^2\}^{\frac{1}{2}}$$

$$= \max\{74,68\}^{1/2} = (74)^{1/2}$$

$$C^*(S\pi,S\sigma) = 8 - (1 + 1 + 1 + 1 + 0 + 0 + |1-2|) = 3.$$

<u>Proof of Theorem</u>: The Hausdorff distances induced by the bi-invariant metrics H and C are derived here, while the proofs for T^*, U^*, F^*, and R^* are deferred to Section D.

<u>Proof for H^*</u>: Since H is left-invariant, $H^*(S\pi,S\sigma) = \min_{\beta\in S\sigma} H(\pi,\beta)$ by the Lemma of Section III.B, pages 21-22.

Now

$$\min_{\beta\varepsilon S\sigma} H(\pi,\beta) = \min_{\beta\varepsilon S\sigma} \#\{j = 1,\ldots,n\colon \pi(j) \neq \beta(j)\}$$

$$= \min_{\beta\varepsilon S\sigma} \#\{j=1,\ldots,n\colon \pi^{-1}(j) \neq \beta^{-1}(j)\}$$

$$= n - \max_{\beta\varepsilon S\sigma} \#\{j=1,\ldots,n\colon \pi^{-1}(j) = \beta^{-1}(j)\}$$

$$= n - \max_{\beta\varepsilon S\sigma} \sum_{i=1}^{r} \#\{j\varepsilon N_i\colon \pi^{-1}(j) = \beta^{-1}(j)\}$$

$$\geq n - \max_{\beta\varepsilon S\sigma} \sum_{i=1}^{r} \#(\pi^{-1}(N_i) \cap \beta^{-1}(N_i))$$

$$= n - \sum_{i=1}^{r} \#(\pi^{-1}(N_i) \cap \sigma^{-1}(N_i)) = n - \sum_{i=1}^{r} n_{ii},$$

so $H^*(S\pi,S\sigma) \geq n - \sum_{i=1}^{r} n_{ii}$. To prove the reverse inequality, for each $i = 1,\ldots,r$, let $p_{i1},p_{i2},\ldots,p_{i,n_i-n_{ii}}$ and $s_{i1},s_{i2},\ldots,s_{i,n_i-n_{ii}}$ be enumerations of the sets $\pi^{-1}(N_i)\cap\sigma^{-1}(N_i^c)$ and $\pi^{-1}(N_i^c)\cap\sigma^{-1}(N_i)$, respectively. Let β_0 be the permutation

$$\beta_0(j) = \pi(j) \qquad \forall j\varepsilon \bigcup_{i=1}^{r} [\pi^{-1}(N_i)\cap\sigma^{-1}(N_i)]$$

$$\beta_0(s_{ij}) = \pi(p_{ij}) \quad \forall i=1,\ldots,r\ \forall j=1,\ldots,n_i-n_{ii}.$$

Then $\beta_0\varepsilon S\sigma$ since $\beta_0^{-1}(N_i) = \sigma^{-1}(N_i)$ for all $i = 1,\ldots,r$, and

$$H(\pi,\beta_0) = n - \#\{j=1,\ldots,n\colon \pi(j) = \beta_0(j)\}$$

$$= n - \#\bigcup_{i=1}^{r} [\pi^{-1}(N_i)\cap\sigma^{-1}(N_i)] = n - \sum_{i=1}^{r} n_{ii}.$$

Thus

$$H^*(S\pi,S\sigma) \leq n - \sum_{i=1}^{r} n_{ii}. \qquad \square$$

<u>Proof for C^*</u>: Since C^* is right-invariant, it is enough to prove the result when π = id. Now

$$C^*(S,S\sigma) = \min_{\beta\varepsilon S\sigma} C(id,\beta) = n - \max_{\beta\varepsilon S\sigma} (\#\text{ cycles in } \beta)$$

where the first equality follows from the left-invariance of C and the Lemma of Section III.B. Let $m = \max_{\beta\varepsilon S\sigma} (\#\text{ cycles in } \beta)$.

<u>r = 2</u>: It remains to show $m = n - n_{12}$. To begin with, observe $n_{12} = n_{21}$, because $n_{11} + n_{12} = n_1 = n_{11} + n_{21}$.

Let $\beta\varepsilon S\sigma$. If β has c one-cycles, the number of cycles of length ≥ 2 is $\leq \frac{n-c}{2}$, so the total number of cycles is $\leq c + \frac{n-c}{2}$. This bound is an increasing function of c, and $c \leq n_{11} + n_{22}$, so the total number of cycles is

$$\begin{aligned} &\leq n_{11} + n_{22} + \frac{n-(n_{11}+n_{22})}{2} \\ &= n_{11} + n_{22} + \frac{n_{12} + n_{21}}{2} \\ &= n_{11} + n_{22} + n_{21} = n-n_{12}. \end{aligned}$$

Hence $m \leq n - n_{12}$.

On the other hand, let $a_1,\ldots,a_{n_{21}}$ and $b_1,\ldots,b_{n_{21}}$ be enumerations of the sets $N_1 \cap \sigma^{-1}(N_2)$ and $N_2 \cap \sigma^{-1}(N_1)$, respectively, and define $\beta_0 \varepsilon S_n$ by

$$\begin{aligned} &\beta_0(j) = j && \forall j \varepsilon \bigcup_{i=1}^{2} (N_i \cap \sigma^{-1}(N_i)) \\ &\beta_0(a_j) = b_j && \forall j = 1,\ldots,n_{21} \\ &\beta_0(b_j) = a_j && \forall j=1,\ldots,n_{21}. \end{aligned}$$

Then $\beta_0 \varepsilon S\sigma$ and has $n_{11} + n_{22}$ one-cycles and n_{21} two-cycles, so $m \geq n_{11} + n_{22} + n_{21} = n - n_{12}$. $\square$

<u>r = 3</u>: It is enough to show

$$m = \sum_{i=1}^{3} n_{ii} + \sum_{i<j} \min \{n_{ij}, n_{ji}\} + |n_{12} - n_{21}|.$$

First note that the equalities

$$n_{12} + n_{22} + n_{32} = n_{21} + n_{22} + n_{23}$$

and

$$n_{11} + n_{12} + n_{13} = n_{11} + n_{21} + n_{31}$$

imply that

$$n_{12} - n_{21} = n_{23} - n_{32} = n_{31} - n_{13}.$$

Let $\Delta = n_{12} - n_{21}$, and let $m_{ij} = \min \{n_{ij}, n_{ji}\}$ for $i \neq j$.

Let $\beta \varepsilon S\sigma$. For $i,j = 1,2,3$, let e_i be the number of one-cycles of β consisting of a number from N_i, and let e_{ij} denote the number of two-cycles of β consisting of one number from N_i and one number from N_j. The total number of cycles is then

$$\leq \sum_{i=1}^{3} e_i + \sum_{i=1}^{3} e_{ii} + \sum_{i<j} e_{ij} + \frac{1}{3}\left(n - \sum_{i=1}^{3} e_i - 2\sum_{i=1}^{3} e_{ii} - 2\sum_{i<j} e_{ij}\right)$$

$$= \frac{2}{3}\sum_{i=1}^{3} e_i + \frac{1}{3}\sum_{i=1}^{3} e_{ii} + \frac{1}{3}\sum_{i<j} e_{ij} + \frac{n}{3}.$$

But $e_{ii} \leq \frac{n_{ii} - e_i}{2}$ and $e_{ij} \leq m_{ij}$, so the number of cycles is

$$\leq \frac{2}{3} \sum_{i=1}^{3} e_i + \frac{1}{3} \sum_{i=1}^{3} \frac{(n_{ii}-e_i)}{2} + \frac{1}{3} \sum_{i<j} m_{ij} + \frac{n}{3}$$

$$= \frac{1}{2} \sum_{i=1}^{3} e_i + \frac{1}{6} \sum_{i=1}^{3} n_{ii} + \frac{1}{3} \sum_{i<j} m_{ij} + \frac{n}{3}.$$

Finally, substituting $e_i \leq n_{ii}$ and

$$n = \sum_{i=1}^{3} n_{ii} + \sum_{i<j} (n_{ij} + n_{ji}) = \sum_{i=1}^{3} n_{ii} + \sum_{i<j} (2m_{ij} + |\Delta|),$$

one obtains the desired bound

$$m \leq \sum_{i=1}^{3} n_{ii} + \sum_{i<j} m_{ij} + |\Delta|.$$

To prove the reverse inequality, for each $i \neq j$ choose $a_1^{ij},\ldots,a_{m_{ij}}^{ij}$ from $N_i \cap \sigma^{-1}(N_j)$. Let $b_1^i,\ldots,b_{|\Delta|}^i$ be an enumeration of the rest of $N_i \cap \sigma^{-1}(N_i^c)$. Define $\beta_0 \varepsilon S_n$ by

$$\beta_0(j) = j \qquad \forall j \in \bigcup_{i=1}^{3} (N_i \cap \sigma^{-1}(N_i))$$

$$\beta_0(a_k^{ij}) = a_k^{ji} \qquad \forall i \neq j \; \forall\, k = 1,\ldots,m_{ij}$$

$$\beta_0(b_k^1) = b_k^2,\ \beta_0(b_k^2) = b_k^3,\ \beta_0(b_k^3) = b_k^1 \quad \forall k=1,\ldots,\Delta \quad \underline{\text{if } \Delta \geq 0}.$$

$$\beta_0(b_k^1) = b_k^3,\ \beta_0(b_k^2) = b_k^1,\ \beta_0(b_k^3) = b_k^2 \quad \forall k=1,\ldots,|\Delta| \quad \underline{\text{if } \Delta < 0}.$$

Then $\beta_0 \varepsilon S\sigma$ and has $\sum_{i=1}^{3} n_{ii}$ one-cycles, $\sum_{i<j} m_{ij}$ two-cycles, and $|\Delta|$ three-cycles, so $m \geq \sum_{i=1}^{3} n_{ii} + \sum_{i<j} m_{ij} + |\Delta|$. □

When $r > 3$, the author has been unable to obtain an explicit formula for C^*, beyond the fact that

$$C^*(S\pi, S\sigma) = \min_{\beta \in S\sigma} C(\pi, \beta).$$

C. The Fixed Vector Metrics on S_n/S

The construction of Section III.C can be carried over to the present situation, and provides additional metrics on the coset space S_n/S.

In the notation of III.C, take $G = S_n$, $K = S$, ρ = the n-dimensional permutation representation, and let $\|\cdot\|$ be the ℓ^1 norm on $\mathbb{R}^n$. As remarked in III.C, $\rho(\pi)^T$ is "norm-preserving," because it simply permutes the components of any vector $\underset{\sim}{w}$:

$$(*) \qquad (\rho(\pi)^T \underset{\sim}{w})_i = w_{\pi(i)} \quad \forall \pi \in S_n \quad \forall \underset{\sim}{w} \in \mathbb{R}^n.$$

To complete the construction of the fixed vector metric, we must specify a vector $\underset{\sim}{v} \in \mathbb{R}^n$ which is fixed by S under ρ^T, but which is not fixed by any larger subgroup. The general question as to when such fixed vectors exist is addressed in Appendix A; but in the case at hand, it is clear that fixed vectors exist, and there is a variety of choices. Indeed, by (*) above, we may choose any vector $\underset{\sim}{v} = (v_1, \ldots, v_n)^T$ such that

$$v_i = c_j \text{ for all } i \in N_j,$$

where the c_j are distinct real numbers for $j = 1, \ldots, r$.

The fixed vector approach then yields the following metric on S_n/S:

$$\begin{aligned} F_{fv,\underset{\sim}{v}}(S\pi,S\sigma) &= \|(\rho(\pi)-\rho(\sigma))^T \underset{\sim}{v}\|_1 \\ &= \sum_{i=1}^{n} |(\rho(\pi)^T \underset{\sim}{v})_i - (\rho(\sigma)^T \underset{\sim}{v})_i| \\ &= \sum_{i=1}^{n} |v_{\pi(i)} - v_{\sigma(i)}| \\ &= \sum_{i=1}^{r} \sum_{j=1}^{r} n_{ij}|c_i - c_j|. \end{aligned}$$

As explained in Section III.C, this can be regarded as an extension of Spearman's footrule (since it _is_ the footrule when $K = \{id\}$ and $v_i = i$). It has the interpretation that the two judges are assigning the "pseudo-ranks" c_i and c_j, respectively, to those items in $\pi^{-1}(N_i) \cap \sigma^{-1}(N_j)$, and then computing Spearman's footrule using the pseudo-ranks rather than the ordinary ranks. This is especially meaningful when for all $i = 1,\ldots,r$, the pseudo-rank

$$c_i = \left(\sum_{j=1}^{i-1} n_j\right) + \frac{n_i+1}{2}$$

is the _average_ of the ranks in the set N_i. The fixed vector metric is denoted simply F_{fv} in this case.

If instead we take $\|\cdot\|$ to be the ℓ^2 norm on $\mathbb{R}^n$, we obtain the following extension of Spearman's _rho_ to S_n/S:

$$R_{fv,\underset{\sim}{v}}(S\pi, S\sigma) = \left[\sum_{i=1}^{r} \sum_{j=1}^{r} n_{ij}(c_i - c_j)^2\right]^{\frac{1}{2}}$$

which has a similar interpretation in terms of "pseudo-ranks."

D. Hausdorff Distances between Different Types of Partially Ranked Data: A Complete Proof of the Main Theorem

In this section, the problem of computing a nonparametric measure of association is considered for the most general situation, where the two partial rankings are of <u>different</u> types. The partial rankings then correspond to right cosets of <u>different</u> subgroups of S_n, but the Hausdorff distance between them can still be calculated.

Examples of such data typically arise in certain elections, such as those of the American Psychological Association and the American Mathematical Society, and the Australian governmental elections. Voters are asked to rank the candidates, up to the point where they have no preference among those who remain. Thus, one voter might list his five favorite candidates out of ten, while a second voter might specify only his first, second, and third choices. The resulting partial rankings would then be of type 1,1,1,1,1,5 and of type 1,1,1,7, respectively.

To study the problem in a general context, suppose the two partial rankings are of types $n_1,\ldots,n_r$ and $n_1',\ldots,n_{r'}'$, respectively. Let $\pi,\sigma\varepsilon S_n$ be any full rankings which induce the given partial rankings. Then, as shown in Section A, the set of all full rankings which induce the first partial ranking is the right coset $S\pi$, where $S = S_{n_1}\times\ldots\times S_{n_r}\subset S_n$; while the set of full rankings which induce the second partial ranking is the right coset $S'\sigma$, where $S' = S_{n_1'}\times\ldots\times S_{n_{r'}'}\subset S_n$.

To measure the agreement of the partial rankings, we can still compute the Hausdorff distance

$$d^*(S\pi,S'\sigma) = \max\{\max_{\beta\varepsilon S'\sigma}\ \min_{\alpha\varepsilon S\pi} d(\alpha,\beta), \max_{\alpha\varepsilon S\pi}\ \min_{\beta\varepsilon S'\sigma} d(\alpha,\beta)\}.$$

Indeed, if X is any set endowed with a bounded metric d, then the induced Hausdorff distance d^* is well-defined between any two closed non-empty subsets of X, and satisfies the axioms for a metric on such subsets of X [N1,D4,K7,K8]. In the case at hand, $S\pi$ and $S'\sigma$ are subsets of the <u>finite</u> metric space (S_n,d), so $d^*(S\pi,S'\sigma)$ makes sense. Furthermore, it is easy to see that if d is right-invariant, then so is d^*; the importance of this was discussed in II.C.

However, there <u>is</u> a price to be paid for the loss of some group-theoretic structure, in the sense that the <u>calculation</u> of d^* is frequently more difficult when the two partial rankings are of different types. For example, the Lemma of Section III.B, pages 21-22 is no longer applicable.

Nonetheless, these Hausdorff distances will now be derived, in what can be considered the "main theorem" of this monograph, since most of the earlier results are direct consequences of it.

In the course of the proof, it will be seen that three of the metrics on S_n —— T, F, and R —— satisfy a special property called the "transposition property." This is described in Lemma 2. The transposition property is of some interest in its own right, and it simplifies the calculation of the induced Hausdorff distances.

Before stating the main theorem, we generalize our previous notation slightly, to take care of the present situation.

Notation: Let $N_1,\ldots,N_r$ be the partition of $\{1,\ldots,n\}$ given in Section A, and let $N_1',\ldots,N_{r'}'$ be the partition

$$N_1' = \{1,2,\ldots,n_1'\}$$

$$N_2' = \{n_1'+1,\ldots,n_1'+n_2'\}$$

$$\vdots$$

$$N_{r'}' = \{n_1'+\ldots+n_{r'-1}'+1,\ldots,n\}.$$

For $i = 1,\ldots,r$ and $j = 1,\ldots,r'$, let

$$n_{ij} = \#[\pi^{-1}(N_i)\cap\sigma^{-1}(N_j')],$$

so, as in Section B, n_{ij} is the number of items placed in the i-th category by the first judge and in the j-th category by the second.

Theorem: The Hausdorff distances between two partial rankings of types $n_1,\ldots,n_r$ and $n_1',\ldots,n_{r'}'$, induced by the six metrics T, F, R, U, H, and C of Section II.B, are:

$$T^*(S\pi,S'\sigma) = \max\{\sum_{\substack{i<i'\\ j\geq j'}} n_{ij}n_{i'j'},\ \sum_{\substack{i\leq i'\\ j>j'}} n_{ij}n_{i'j'}\}$$

$$F^*(S\pi,S'\sigma) = \max\{\sum_{i=1}^{r}\sum_{j=1}^{r'} n_{ij}\left|\sum_{k=1}^{i-1} n_k - \sum_{k=1}^{j-1} n_k' + \sum_{k=1}^{j-1} n_{ik} - \sum_{k=i+1}^{r} n_{kj}\right|,$$

$$\sum_{i=1}^{r}\sum_{j=1}^{r'} n_{ij}\left|\sum_{k=1}^{i-1} n_k - \sum_{k=1}^{j-1} n_k' + \sum_{k=j+1}^{r'} n_{ik} - \sum_{k=1}^{i-1} n_{kj}\right|\}$$

$$R^*(S\pi,S'\sigma) = \max\{\sum_{i=1}^{r}\sum_{j=1}^{r'} n_{ij}\left(\sum_{k=1}^{i-1} n_k - \sum_{k=1}^{j-1} n_k' + \sum_{k=1}^{j-1} n_{ik} - \sum_{k=i+1}^{r} n_{kj}\right)^2,$$

$$\sum_{i=1}^{r}\sum_{j=1}^{r'} n_{ij}\left(\sum_{k=1}^{i-1} n_k - \sum_{k=1}^{j-1} n_k' + \sum_{k=j+1}^{r'} n_{ik} - \sum_{k=1}^{i-1} n_{kj}\right)^2\}^{\frac{1}{2}}$$

$$U^*(S\pi,S'\sigma) = n\text{-min}\left\{ \max_{\substack{i(j)\text{ a non-decreasing}\\ \text{function of } j}} \sum_{j=1}^{r'} n_{i(j)j},\ \max_{\substack{j(i)\text{ a non-decreasing}\\ \text{function of } i}} \sum_{i=1}^{r} n_{ij(i)} \right\}$$

$$H^*(S\pi,S'\sigma) = n\text{-min}\left\{ \sum_{i=1}^{r}\sum_{j=1}^{r'} \max\{0,n_{ij}+\nu_{ij}-n_j'\},\ \sum_{i=1}^{r}\sum_{j=1}^{r'} \max\{0,n_{ij}+v_{ij}-n_i\}\right\}$$

where $\nu_{ij} = \#(N_i \cap N_j')$ for $i = 1,\dots,r$ and $j = 1,\dots,r'$.

Finally, for the special case in which the first judge lists his top k and the second judge lists his top k' choices, so $S = S_{n-k} = \{\pi\varepsilon S_n\colon \pi(i) = i\ \forall i=1,\dots,k\}$ and $S' = S_{n-k'} = \{\pi\varepsilon S_n\colon \pi(i) = i\ \forall i=1,\dots k'\}$:

$$C^*(S_{n-k}\pi,S_{n-k'}\sigma) = k-\left(\begin{matrix}\text{number of cycles in } \sigma\pi^{-1}\\ \text{consisting entirely of numbers} \le k'\end{matrix}\right) \quad \text{if } n > k \ge k';$$

$$C^*(S_{n-k}\pi,S_{n-k'}\sigma) = n-1-\left(\begin{matrix}\text{number of cycles in } \sigma\pi^{-1} \text{ con-}\\ \text{sisting entirely of numbers} \le k'\end{matrix}\right) \quad \text{if } n = k > k'.$$

<u>Proof</u>: The proof relies on a series of lemmas, the first of which describes a handy way of computing the Hausdorff distance.

<u>Lemma 1</u>: Let d be a metric on S_n. For any $\sigma\varepsilon S_n$, let $\pi_{min}(\sigma)$ minimize $d(\cdot,\sigma)$, subject to the constraint $\pi_{min}(\sigma)\varepsilon S\pi$, and let σ_{max} maximize $d(\pi_{min}(\cdot),\cdot)$, subject to the constraint $\sigma_{max}\varepsilon S'\sigma$. Similarly, for any $\pi\varepsilon S_n$, let $\sigma_{min}(\pi)$ minimize $d(\pi,\cdot)$, subject to the constraint $\sigma_{min}(\pi)\varepsilon S'\sigma$,

and let π_{max} maximize $d(\cdot,\sigma_{min}(\cdot))$, subject to the constraint $\pi_{max} \varepsilon S\pi$. Then:

$$d^*(S\pi,S'\sigma) = \max\{d(\pi_{min}(\sigma_{max}),\sigma_{max}),\ d(\sigma_{min}(\pi_{max}),\pi_{max})\}.$$

<u>Proof of Lemma 1</u>:

$$d^*(S\pi,S'\sigma) = \max\{\max_{\beta\varepsilon S'\sigma}\ \min_{\alpha\varepsilon S\pi} d(\alpha,\beta),\ \max_{\alpha\varepsilon S\pi}\ \min_{\beta\varepsilon S'\sigma} d(\alpha,\beta)\}$$

$$= \max\{\max_{\beta\varepsilon S'\sigma} d(\pi_{min}(\beta),\beta),\ \max_{\alpha\varepsilon S\pi} d(\alpha,\sigma_{min}(\alpha))\}$$

$$= \max\{d(\pi_{min}(\sigma_{max}),\sigma_{max}),\ d(\pi_{max},\sigma_{min}(\pi_{max}))\}. \ \square$$

The next order of business is to find σ_{max}, $\pi_{min}(\sigma_{max})$, π_{max}, and $\sigma_{min}(\pi_{max})$. It turns out that these calculations are very much alike when d is any of the metrics T, F, R, or U. In the case of the metrics T, F, and R, the reason for this similarity lies essentially in the fact that they share the following "transposition property."

<u>Lemma 2</u>: Let d = T, F, or R. Let $\alpha,\beta,\gamma\varepsilon S_n$ be permutations such that α and β differ by a single transposition; that is, there exist integers $p,q\varepsilon\{1,\dots,n\}$ such that

$$\alpha(p) = \beta(q)$$

$$\alpha(q) = \beta(p)$$

$$\alpha(i) = \beta(i) \quad \forall i \neq p,q.$$

Suppose $\alpha(p) \leq \alpha(q)$ and $\gamma(p) \leq \gamma(q)$. Then $d(\alpha,\gamma) \leq d(\beta,\gamma)$.

<u>Remark</u>: The lemma is not true when d = U, H, or C. Indeed, in the bracket notation of Section II.A, page 5, let $\alpha,\beta,\gamma\varepsilon S_4$ be given by:

$$\alpha = <2,1,4,3>, \qquad \beta = <4,1,2,3>, \qquad \gamma = \text{id},$$

and let p = 2, q = 4. The hypotheses of the lemma are satisfied, but $U(\alpha,\gamma) = 2 > 1 = U(\beta,\gamma)$.

For a counterexample when d = H or C, let

$$\alpha = <1,4,2,3>, \qquad \beta = <1,4,3,2>, \quad \gamma = \text{id}, \quad p = 2, \quad q = 3.$$

The hypotheses of the lemma are satisfied, but

$$H(\alpha,\gamma) = 3 > 2 = H(\beta,\gamma)$$

and

$$C(\alpha,\gamma) = 2 > 1 = C(\beta,\gamma).$$

<u>Proof of Lemma 2</u>: <u>d = T</u>: Recall that $T(\alpha,\gamma)$ is the number of pairs (i,j) such that $\alpha(i) < \alpha(j)$ and $\gamma(i) > \gamma(j)$. Therefore, if (i,j) contributes to 1 to $T(\alpha,\gamma)$ but contributes 0 to $T(\beta,\gamma)$, then

(*) $\qquad \alpha(i) < \alpha(j),\ \beta(i) > \beta(j), \text{ and } \gamma(i) > \gamma(j).$

Since $\alpha(k) = \beta(k) \ \ \forall k \notin \{p,q\}$, there are only two possible ways for (*) to hold: either i = p and $\alpha(p) < \alpha(j) < \alpha(q)$, or j = q and $\alpha(p) < \alpha(i) < \alpha(q)$. In the first case, one has $\alpha(q) > \alpha(j)$, $\beta(q) = \alpha(p) < \alpha(j) = \beta(j)$, and $\gamma(q) \geq \gamma(p) = \gamma(i) > \gamma(j)$, so the contribution from (p,j) to $T(\alpha,\gamma)$ is matched by the contribution from (q,j) to $T(\beta,\gamma)$. In the second case, one has $\alpha(i) > \alpha(p)$, $\beta(i) = \alpha(i) < \alpha(q) = \beta(p)$, and $\gamma(i) > \gamma(j) = \gamma(q) \geq \gamma(p)$, so the contribution from (i,q) to $T(\alpha,\gamma)$ is matched by the contribution from (i,p) to $T(\beta,\gamma)$.

<u>d = F</u>:

$$F(\beta,\gamma)-F(\alpha,\gamma) = |\beta(p)-\gamma(p)|+|\beta(q)-\gamma(q)|-|\alpha(p)-\gamma(p)|-|\alpha(q)-\gamma(q)|$$

$$= |\alpha(q)-\gamma(p)|+|\alpha(p)-\gamma(q)|-|\alpha(p)-\gamma(p)|-|\alpha(q)-\gamma(q)|.$$

If $\alpha(p) \geq \gamma(q)$ then $\alpha(q) \geq \alpha(p) \geq \gamma(q) \geq \gamma(p)$, and the above sum is

$$(\alpha(q)-\gamma(p))+(\alpha(p)-\gamma(q))-(\alpha(p)-\gamma(p))-(\alpha(q)-\gamma(q)) = 0.$$

Similarly, if $\gamma(p) \geq \alpha(q)$, then $\gamma(q) \geq \gamma(p) \geq \alpha(q) \geq \alpha(p)$, and $F(\beta,\gamma)-F(\alpha,\gamma) = 0$. Finally, suppose $\alpha(p) < \gamma(q)$ and $\gamma(p) < \alpha(q)$. Then

$$F(\beta,\gamma)-F(\alpha,\gamma) = \alpha(q)-\gamma(p)+\gamma(q)-\alpha(p)-|\alpha(p)-\gamma(p)|-|\alpha(q)-\gamma(q)|$$

$$= \alpha(q)+\gamma(q)-|\alpha(q)-\gamma(q)|-(\alpha(p)+\gamma(p)+|\alpha(p)-\gamma(p)|)$$

$$= 2 \min \{\alpha(q),\gamma(q)\} - 2 \max \{\alpha(p),\gamma(p)\} \geq 0.$$

<u>d=R</u>: First observe that if x, y, and ε are real numbers satisfying $x-y \geq \varepsilon \geq 0$, then

$$x^2 + y^2 \geq (x-\varepsilon)^2 + (y+\varepsilon)^2.$$

Indeed, $x^2 + y^2 - (x-\varepsilon)^2 - (y+\varepsilon)^2 = -2\varepsilon^2 + 2x\varepsilon - 2y\varepsilon = 2\varepsilon(x-y-\varepsilon) \geq 0$.

Now

$$R^2(\beta,\gamma)-R^2(\alpha,\gamma)= (\beta(p)-\gamma(p))^2 + (\beta(q)-\gamma(q))^2$$

$$- (\alpha(p)-\gamma(p))^2 - (\alpha(q)-\gamma(q))^2$$

$$= (\alpha(q)-\gamma(p))^2 + (\alpha(p)-\gamma(q))^2$$

$$- (\alpha(p)-\gamma(p))^2 - (\alpha(q)-\gamma(q))^2.$$

It remains to show this is non-negative. If $\alpha(p) \geq \gamma(q)$ then $\alpha(q) \geq \alpha(p) \geq \gamma(q) \geq \gamma(p)$, and the preceding observation applies with $x = \alpha(q)-\gamma(p)$, $y = \alpha(p)-\gamma(q)$, and $\varepsilon = \alpha(q)-\alpha(p)$. Similarly, if $\gamma(p) \geq \alpha(q)$ then $\gamma(q) \geq \gamma(p) \geq \alpha(q) \geq \alpha(p)$, and the preceding observation applies with $x = \gamma(q)-\alpha(p)$, $y = \gamma(p)-\alpha(q)$, and $\varepsilon = \alpha(q)-\alpha(p)$. Finally, suppose $\alpha(p) < \gamma(q)$ and $\gamma(p) < \alpha(q)$, and let

$$M_q = \max\{\alpha(q),\gamma(q)\},\quad m_q = \min\{\alpha(q),\gamma(q)\},$$

$$M_p = \max\{\alpha(p),\gamma(p)\},\quad m_p = \min\{\alpha(p),\gamma(p)\}.$$

Then $M_q \geq m_q \geq M_p \geq m_p$, and $R^2(\beta,\gamma) - R^2(\alpha,\gamma)$ is either

$$(M_q-m_p)^2 + (m_q-M_p)^2 - (M_p-m_p)^2 - (M_q-m_q)^2$$

or

$$(M_q-M_p)^2 + (m_q-m_p)^2 - (M_p-m_p)^2 - (M_q-m_q)^2.$$

The first expression is non-negative since $|M_q-m_p| \geq |M_q-m_q| + |M_p-m_p|$, while the second is non-negative since $|M_q-M_p| \geq |M_q-m_q|$ and $|m_q-m_p| \geq |M_p-m_p|$. □

The importance of the preceding lemma lies in the fact that when the "transposition property" holds, the permutations σ_{max}, $\pi_{min}(\sigma_{max})$, π_{max}, and $\sigma_{min}(\pi_{max})$ have a very special form.

<u>Lemma 3</u>: Let d = T, F, R, or, more generally, any metric for which Lemma 2 holds. For $i = 1,\ldots,r$ and $j = 1,\ldots,r'$, let N_{ij} denote the n_{ij} numbers in the set $\pi^{-1}(N_i)\cap\sigma^{-1}(N_j')$, arranged in ascending order (or in any fixed order, for that matter). Then

$$\sigma_{max} = \langle N_{r1},\ldots,N_{11},N_{r2},\ldots,N_{12},\ldots,N_{rj},\ldots,N_{1j},\ldots,N_{rr'},\ldots,N_{1r'}\rangle$$

$$\pi_{min}(\sigma_{max}) = \langle N_{11},\ldots,N_{1r'},\ldots,N_{i1},\ldots,N_{ir'},\ldots,N_{r1},\ldots,N_{rr'}\rangle$$

$$\pi_{max} = \langle N_{1r'},\ldots,N_{11},\ldots,N_{ir'},\ldots,N_{i1},\ldots,N_{rr'},\ldots,N_{r1}\rangle$$

$$\sigma_{min}(\pi_{max}) = \langle N_{11},\ldots,N_{r1},\ldots,N_{1j},\ldots,N_{rj},\ldots,N_{1r'},\ldots,N_{rr'}\rangle.$$

<u>Proof of Lemma 3</u>: First of all, it is claimed that for any $\sigma_0 \varepsilon S'\sigma$,

(*) $\pi_{min}(\sigma_0) = \langle N_{11}(\sigma_0),\ldots,N_{1r'}(\sigma_0),\ldots,N_{r1}(\sigma_0),\ldots,N_{rr'}(\sigma_0)\rangle$,

where $N_{ij}(\sigma_0)$ denotes the n_{ij} numbers in the set $\pi^{-1}(N_i) \cap \sigma^{-1}(N_j')$, arranged in the same order in which they occur in the bracket-notation version of σ_0. To see this, let $\pi_0 \varepsilon S\pi$ be arbitrary, and let π_1 denote the right hand side of (*); it is sufficient to show that π_1 can be obtained from π_0 by a sequence of transpositions which satisfy the hypotheses of Lemma 2. Starting off with π_0, the first transposition switches $\pi_0^{-1}(1)$ and $\pi_1^{-1}(1)$ (in the bracket-notation version); call the resulting permutation $\pi_{(1)}$. The second transposition switches $\pi_{(1)}^{-1}(2)$ and $\pi_1^{-1}(2)$, and so on; the k-th transposition switches $\pi_{(k-1)}^{-1}(k)$ and $\pi_1^{-1}(k)$, where $\pi_{(k-1)}$ is the result of the first k-1 transpositions. The k-th transposition satisfies the hypotheses of Lemma 2, with $\beta = \pi_{(k-1)}$, $\alpha = \pi_{(k)}$, $\gamma = \sigma_0$, $p = \pi_1^{-1}(k)$, and $q = \pi_{(k-1)}^{-1}(k)$. Indeed, an easy induction on m shows that

(**) $\pi_{(m)}(\pi_1^{-1}(i)) = i = \pi_1(\pi_{(m)}^{-1}(i)) \quad \forall i \le m \quad \forall m = 1,\ldots,n-1$,

and hence

$$\alpha(q) = \beta(p) = \pi_{(k-1)}(\pi_1^{-1}(k)) \geq k = \beta(q) = \alpha(p).$$

On the other hand, $\gamma(q) \geq \gamma(p)$, since $p,q \varepsilon \bigcup_{j=1}^{r'} N_{ij}$ for some i, the numbers in $\bigcup_{j=1}^{r'} N_{ij}$ occur in the same order in σ_0 and π_1, and

$$\pi_1(q) = \pi_1(\pi_{(k-1)}^{-1}(k)) \geq k = \pi_1(p).$$

Hence the hypotheses of Lemma 2 are satisfied, so

$$d(\pi_{(k)},\sigma_0) \leq d(\pi_{(k-1)},\sigma_0) \leq \ldots \leq d(\pi_0,\sigma_0).$$

The proof of (*) is complete, since $\pi_{(n-1)} = \pi_1$ by (**).

One upshot of (*) and the right-invariance of d is that $d(\pi_{min}(\sigma_0),\sigma_0)$ does not depend on the relative order in which the elements of N_{ij} appear in the bracket-notation version of σ_0, because these elements appear in the same order in σ_0 and $\pi_{min}(\sigma_0)$. Hence, for the purpose of calculating σ_{max}, one may as well assume that these numbers occur in ascending order in both σ_{max} and $\pi_{min}(\sigma_{max})$, and compute σ_{max} subject to these constraints. Thus, let σ_1 be the version of σ_{max} proposed by the lemma, let

$$\pi_0 = \pi_{min}(\sigma_1) = \langle N_{11},\ldots,N_{1r'},N_{21},\ldots,N_{2r'},\ldots,N_{r1},\ldots,N_{rr'}\rangle,$$

and let $\sigma_0 \varepsilon S'\sigma$ be arbitrary, subject to the constraints that for all $i = 1,\ldots,r$ and $j = 1,\ldots,r'$, the numbers in N_{ij} appear in ascending order in σ_0. To prove that $\sigma_1 = \sigma_{max}$, it remains to show that $d(\sigma_1,\pi_0) \geq d(\sigma_0,\pi_0)$.

Just as before, this is accomplished by obtaining σ_1 from σ_0 by a sequence of transpositions which satisfy the hypotheses of Lemma 2.

Starting off with σ_0, the first transposition switches $\sigma_0^{-1}(1)$ and $\sigma_1^{-1}(1)$; call the resulting permutation $\sigma_{(1)}$. The second transposition switches $\sigma_{(1)}^{-1}(2)$ and $\sigma_1^{-1}(2)$, and so on; the k-th transposition switches $\sigma_{(k-1)}^{-1}(k)$ and $\sigma_1^{-1}(k)$, where $\sigma_{(k-1)}$ is the result of the first k-1 transpositions. The k-th transposition satisfies the hypotheses of Lemma 2, with $\alpha = \sigma_{(k-1)}$, $\beta = \sigma_{(k)}$, $\gamma = \pi_0$, $p = \sigma_{(k-1)}^{-1}(k)$, and $q = \sigma_1^{-1}(k)$. The proof follows the same lines as before, except for the verification that $\gamma(q) \geq \gamma(p)$: if $p \neq q$, the additional "ascending order constraint" implies that $p \varepsilon N_{ij}, q \varepsilon N_{i'j}$ for some $i \neq i'$ and some j. Thus p and q must occur in reverse order in π_0 and σ_1, but just as before, one can show $\sigma_1(p) \geq \sigma_1(q)$.

Hence $d(\sigma_{(k)},\pi_0) \geq d(\sigma_{(k-1)},\pi_0) \geq \ldots \geq d(\sigma_0,\pi_0)$, but $\sigma_{(n-1)} = \sigma_1$, so $d(\sigma_1,\pi_0) \geq d(\sigma_0,\pi_0)$. Thus $\sigma_1 = \sigma_{max}$.

The derivation of $\sigma_{min}(\cdot)$ and π_{max} is completely analogous to that of $\pi_{min}(\cdot)$ and σ_{max}. □

Although Lemma 2 does not hold for Ulam's metric, the conclusion of Lemma 3 does. This is best proved by a direct argument.

<u>Lemma 4</u>: σ_{max}, $\pi_{min}(\sigma_{max})$, π_{max}, and $\sigma_{min}(\pi_{max})$ are the same for Ulam's metric as for T, F, and R, i.e., the same as in the statement of Lemma 3.

<u>Proof of Lemma 4</u>: For any $\sigma_0 \varepsilon S'\sigma$, it is claimed that

(*) $\pi_{min}(\sigma_0) = \langle N_{11}(\sigma_0),\ldots,N_{1r'}(\sigma_0),\ldots,N_{r1}(\sigma_0),\ldots,N_{rr'}(\sigma_0)\rangle$, where the $N_{ij}(\sigma_0)$ are as in the proof of Lemma 3. Indeed, let π_1 be the right hand side of (*), let $\pi_0 \varepsilon S\pi$ be arbitrary, and let $a_1,\ldots,a_m$ be a sequence of maximal length such that

$$\pi_0(a_1) < \ldots < \pi_0(a_m)$$

and

$$\sigma_0(a_1) < \ldots < \sigma_0(a_m).$$

To prove (*), it is enough to show

$$\pi_1(a_k) < \pi_1(a_{k+1}) \quad \forall k = 1,\ldots,m-1.$$

If a_k, a_{k+1} lie in the same N_{ij}, then $\sigma_0(a_k) < \sigma_0(a_{k+1})$ implies $\pi_1(a_k) < \pi_1(a_{k+1})$, by the definitions of π_1 and $N_{ij}(\sigma_0)$. If $a_k \varepsilon N_{ij}$, $a_{k+1} \varepsilon N_{ij'}$ for some $j \neq j'$, then $\sigma_0(a_k) < \sigma_0(a_{k+1})$ implies $j < j'$, and hence $\pi_1(a_k) < \pi_1(a_{k+1})$. Finally, if $a_k \varepsilon N_{ij}$, $a_{k+1} \varepsilon N_{i'j'}$ for some $i \neq i'$, then $\pi_0(a_k) < \pi_0(a_{k+1})$ implies $i < i'$, and therefore $\pi_1(a_k) < \pi_1(a_{k+1})$. Thus (*) is proved.

As in the proof of Lemma 3, we can now assume that the elements of N_{ij} occur in ascending order in both σ_{max} and $\pi_{min}(\sigma_{max})$, and compute σ_{max} subject to these constraints. As before, let σ_1 be the version of σ_{max} proposed by the lemma, let

$$\pi_0 = \pi_{min}(\sigma_1) = \langle N_{11},\ldots,N_{1r'},N_{21},\ldots,N_{2r'},\ldots,N_{r1},\ldots,N_{rr'}\rangle,$$

and let $\sigma_0 \varepsilon S'\sigma$ be arbitrary, subject to the "ascending order constraint." Let $b_1,\ldots,b_p$ be a maximal-length sequence such that

$$\pi_0(b_1) < \ldots < \pi_0(b_p)$$

and

$$\sigma_1(b_1) < \ldots < \sigma_1(b_p).$$

To prove $\sigma_1 = \sigma_{max}$, it is sufficient to show

$$\sigma_0(b_k) < \sigma_0(b_{k+1}) \quad \forall k = 1,\ldots,p-1.$$

If b_k, b_{k+1} lie in the same N_{ij}, then $\sigma_1(b_k) < \sigma_1(b_{k+1})$ implies $\sigma_0(b_k) < \sigma_0(b_{k+1})$, by the "ascending order constraint." It is impossible to have $b_k \varepsilon N_{ij}$ and $b_{k+1} \varepsilon N_{i'j}$ for some $i \neq i'$, because then b_k and b_{k+1} must occur in reverse order in π_0 and σ_1. Finally, if $b_k \varepsilon N_{ij}$ and $b_{k+1} \varepsilon N_{i'j'}$ for some $j \neq j'$, then $\sigma_1(b_k) < \sigma_1(b_{k+1})$ implies $j < j'$, and hence $\sigma_0(b_k) < \sigma_0(b_{k+1})$. Thus $\sigma_1 = \sigma_{max}$.

The proofs for $\sigma_{min}(\cdot)$ and π_{max} are entirely analogous. □

Armed with the machinery of the preceding lemmas, it becomes an easy task to compute T^*, F^*, R^*, and U^*.

<u>Proof of the Theorem for T^*</u>: Recall that $T(\alpha,\beta)$ is the number of pairs (p,q) such that $\alpha(p) < \alpha(q)$ and $\beta(p) > \beta(q)$. By Lemma 3, the contribution to $T(\pi_{min}(\sigma_{max}),\sigma_{max})$ from $N_{ij} \times N_{i'j'}$ is zero unless $i < i'$ and $j \geq j'$, in which case it is $(\#N_{ij})(\#N_{i'j'}) = n_{ij}n_{i'j'}$. Thus

$$T(\pi_{min}(\sigma_{max}),\sigma_{max}) = \sum_{\substack{i<i' \\ j\geq j'}} n_{ij}n_{i'j'}.$$

Similarly, the contribution to $T(\pi_{max},\sigma_{min}(\pi_{max}))$ from $N_{ij} \times N_{i'j'}$ is zero unless $i \leq i'$ and $j > j'$, in which case it is $n_{ij}n_{i'j'}$. Hence

$$T(\pi_{max},\sigma_{min}(\pi_{max})) = \sum_{\substack{i \leq i' \\ j > j'}} n_{ij}n_{i'j'}.$$

By Lemma 1,

$$T^*(S\pi,S'\sigma) = \max \{T(\pi_{min}(\sigma_{max}),\sigma_{max}),\ T(\sigma_{min}(\pi_{max}),\pi_{max})\}$$

$$= \max \{ \sum_{\substack{i < i' \\ j \geq j'}} n_{ij}n_{i'j'},\ \sum_{\substack{i \leq i' \\ j > j'}} n_{ij}n_{i'j'}\}. \quad \square$$

<u>Proof of the Theorem for F^*</u>: Let m_{ij} be the smallest number in N_{ij}. By Lemma 3, there are $\sum_{k=1}^{i-1} n_k + \sum_{k=1}^{j-1} n_{ik}$ numbers occurring to the left of m_{ij} in the bracket representation of $\pi_{min}(\sigma_{max})$, and

$$\sum_{k=1}^{j-1} n_k' + \sum_{k=i+1}^{r} n_{kj}$$

numbers occurring to the left of m_{ij} in the bracket representation of σ_{max}. Hence

$$|\pi_{min}(\sigma_{max})(m_{ij}) - \sigma_{max}(m_{ij})| = |\sum_{k=1}^{i-1} n_k + \sum_{k=1}^{j-1} n_{ik} - \sum_{k=1}^{j-1} n_k' - \sum_{k=i+1}^{r} n_{kj}|.$$

But $|\pi_{min}(\sigma_{max})(m) - \sigma_{max}(m)|$ is constant for $m \in N_{ij}$, so

$$F(\pi_{min}(\sigma_{max}),\sigma_{max}) = \sum_{i=1}^{r} \sum_{j=1}^{r'} n_{ij}|\sum_{k=1}^{i-1} n_k + \sum_{k=1}^{j-1} n_{ik} - \sum_{k=1}^{j-1} n_k' - \sum_{k=i+1}^{r} n_{kj}|.$$

Similarly, there are $\sum_{k=1}^{i-1} n_k + \sum_{k=j+1}^{r'} n_{ik}$ numbers occurring to the left of m_{ij} in the bracket representation of π_{max}, and $\sum_{k=1}^{j-1} n_k' + \sum_{k=1}^{i-1} n_{kj}$ numbers occurring to the left of m_{ij} in the bracket representation of $\sigma_{min}(\pi_{max})$. Hence

$$|\sigma_{min}(\pi_{max})(m_{ij}) - \pi_{max}(m_{ij})| = |\sum_{k=1}^{i-1} n_k + \sum_{k=j+1}^{r'} n_{ik} - \sum_{k=1}^{j-1} n_k' - \sum_{k=1}^{i-1} n_{kj}|.$$

But $|\sigma_{min}(\pi_{max})(m) - \pi_{max}(m)|$ is constant for $m \varepsilon N_{ij}$, so

$$F(\sigma_{min}(\pi_{max}),\pi_{max}) = \sum_{i=1}^{r} \sum_{j=1}^{r'} n_{ij}|\sum_{k=1}^{i-1} n_k + \sum_{k=j+1}^{r'} n_{ik} - \sum_{k=1}^{j-1} n_k' - \sum_{k=1}^{i-1} n_{kj}|.$$

The desired result now follows from Lemma 1. □

<u>Proof of the Theorem for R^*</u>: As shown in the preceding proof for F^*,

$$|\pi_{min}(\sigma_{max})(m) - \sigma_{max}(m)|$$

$$= |\sum_{k=1}^{i-1} n_k + \sum_{k=1}^{j-1} n_{ik} - \sum_{k=1}^{j-1} n_k' - \sum_{k=i+1}^{r} n_{kj}| \quad \forall m \varepsilon N_{ij}.$$

Hence

$$R(\pi_{min}(\sigma_{max}),\sigma_{max})$$

$$= [\sum_{i=1}^{r} \sum_{j=1}^{r'} n_{ij} (\sum_{k=1}^{i-1} n_k + \sum_{k=1}^{j-1} n_{ik} - \sum_{k=1}^{j-1} n_k' - \sum_{k=i+1}^{r} n_{kj})^2]^{\frac{1}{2}}.$$

Similarly,

$$|\sigma_{min}(\pi_{max})(m) - \pi_{max}(m)|$$

$$= |\sum_{k=1}^{i-1} n_k + \sum_{k=j+1}^{r'} n_{ik} - \sum_{k=1}^{j-1} n_k' - \sum_{k=1}^{i-1} n_{kj}| \text{ for all } m \varepsilon N_{ij},$$

and thus

$$R(\sigma_{min}(\pi_{max}),\pi_{max}) = [\sum_{i=1}^{r} \sum_{j=1}^{r'} n_{ij}(\sum_{k=1}^{i-1} n_k + \sum_{k=j+1}^{r'} n_{ik} - \sum_{k=1}^{j-1} n_k' - \sum_{k=1}^{i-1} n_{kj})^2]^{\frac{1}{2}}.$$

The result now follows from Lemma 1. □

<u>Proof of the Theorem for U^*</u>: Let $a_1,\ldots,a_m$ be a sequence of maximal length, such that $\pi_{min}(\sigma_{max})(a_1) < \ldots < \pi_{min}(\sigma_{max})(a_m)$ and $\sigma_{max}(a_1) < \ldots < \sigma_{max}(a_m)$. By Lemma 4, the following must hold:

(1) If $\{a_1,\ldots,a_m\}$ contains some member of N_{ij}, then it cannot contain any member of $N_{i'j}$, for $i \neq i'$.

(2) If $\{a_1,\ldots,a_m\}$ contains some member of N_{ij} and some member of $N_{i'j'}$, where $j < j'$, then $i \leq i'$.

By (1) and (2),

$$m \leq \max_{\substack{i(j) \text{ a non-decreasing} \\ \text{function of } j}} \sum_{j=1}^{r'} n_{i(j)j}.$$

On the other hand, if i(j) is any non-decreasing function of j, then the sequence $N_{i(1)1}, N_{i(2)2},\ldots,N_{i(r')r'}$ occurs in the same order in $\pi_{min}(\sigma_{max})$ and σ_{max}, so

$$m \geq \max_{\substack{i(j) \text{ a non-decreasing} \\ \text{function of } j}} \sum_{j=1}^{r'} n_{i(j)j}.$$

Hence

$$U(\pi_{min}(\sigma_{max}),\sigma_{max}) = n - \max_{\substack{i(j) \text{ a non-decreasing} \\ \text{function of } j}} \sum_{j=1}^{r'} n_{i(j)j}.$$

Similarly,

$$U(\sigma_{min}(\pi_{max}),\pi_{max}) = n - \max_{\substack{j(i) \text{ a non-decreasing} \\ \text{function of } i}} \sum_{i=1}^{r} n_{ij(i)}.$$

The result now follows from Lemma 1. $\square$

Since the conclusion of Lemma 3 does not hold for Hamming distance, one must derive its induced Hausdorff metric by a somewhat different argument.

<u>Proof of the Theorem for H^*</u>: To begin with, it is observed that for any $\sigma_0 \varepsilon S'\sigma$,

$$H(\pi_{min}(\sigma_0),\sigma_0) = n - \sum_{i=1}^{r} \#(\pi^{-1}(N_i) \cap \sigma_0^{-1}(N_i)).$$

Indeed,

$$H(\pi_{min}(\sigma_0),\sigma_0) = \min_{\pi_0 \varepsilon S\pi} H(\pi_0,\sigma_0) = \min_{\pi_0 \varepsilon S\pi} H(\sigma_0,\pi_0) = H^*(S\sigma_0,S\pi)$$

by the Lemma of Section III.B, pages 21-22, and

$$H^*(S\sigma_0,S\pi) = n - \sum_{i=1}^{r} \#(\pi^{-1}(N_i) \cap \sigma_0^{-1}(N_i))$$

by the formula for H^* when S = S', proved in Section IV.B.

It follows that

$$H(\pi_{min}(\sigma_{max}),\sigma_{max})$$

$$= \max_{\sigma_0 \varepsilon S'\sigma} \left(n - \sum_{i=1}^{r} \#(\pi^{-1}(N_i) \cap \sigma_0^{-1}(N_i))\right) = n-m,$$

where

$$m = \min_{\sigma_0 \varepsilon S'\sigma} \sum_{i=1}^{r} \#(\pi^{-1}(N_i) \cap \sigma_0^{-1}(N_i))$$

$$= \min_{\sigma_0 \varepsilon S'\sigma} \sum_{i=1}^{r} \sum_{j=1}^{r'} \#(\pi^{-1}(N_i) \cap \sigma^{-1}(N_j') \cap \sigma_0^{-1}(N_i))$$

$$= \min_{\sigma_0 \varepsilon S'\sigma} \sum_{i=1}^{r} \sum_{j=1}^{r'} \#(N_{ij} \cap \sigma_0^{-1}(N_i)).$$

Now for any $\sigma_0 \varepsilon S'\sigma$,

$$N_{ij} \subset \sigma^{-1}(N_j') = \sigma_0^{-1}(N_j') = \sigma_0^{-1}(N_j' \cap N_i) \cup \sigma_0^{-1}(N_j' \cap N_i^c),$$

and $\#N_{ij} = n_{ij}$ while $\#\sigma_0^{-1}(N_j' \cap N_i^c) = n_j' - \nu_{ij}$, so

$$\#(N_{ij} \cap \sigma_0^{-1}(N_i)) = \#(N_{ij} \cap \sigma_0^{-1}(N_j' \cap N_i))$$

$$\geq \max\{0, n_{ij} - (n_j' - \nu_{ij})\}.$$

Hence

$$m \geq \sum_{i=1}^{r} \sum_{j=1}^{r'} \max\{0, n_{ij} + \nu_{ij} - n_j'\}.$$

On the other hand, let $a_1, \ldots, a_{n_{ij}}$ be an enumeration of N_{ij}. Since $N_j' \cap N_i^c$ has $n_j' - \nu_{ij}$ elements, there exists $\sigma_1 \varepsilon S_n$ such that

$$\sigma_1(a_k) \varepsilon N_j' \cap N_i^c \quad \forall k = 1, \ldots, \min\{n_{ij}, n_j' - \nu_{ij}\} \quad \forall i = 1, \ldots, r$$

$$\forall j = 1, \ldots, r',$$

and such that $\sigma_1(N_{ij}) \subset N_j'$ $\forall i=1,\ldots,r$ $\forall j=1,\ldots,r'$.

Furthermore

$$\#(N_{ij} \cap \sigma_1^{-1}(N_i)) \leq \max\{0, n_{ij} - (n_j' - \nu_{ij})\},$$

and $\sigma_1 \varepsilon S'\sigma$ since $\sigma_1(N_{ij}) \subset N_j'$. Thus

$$m \leq \sum_{i=1}^{r} \sum_{j=1}^{r'} \max\{0, n_{ij} + \nu_{ij} - n_j'\},$$

and

$$H(\pi_{min}(\sigma_{max}),\sigma_{max}) = n - \sum_{i=1}^{r} \sum_{j=1}^{r'} \max\{0, n_{ij} + \nu_{ij} - n_j'\}.$$

A symmetrical argument shows that

$$H(\sigma_{min}(\pi_{max}),\pi_{max}) = n - \sum_{i=1}^{r} \sum_{j=1}^{r'} \max\{0, n_{ij} + \nu_{ij} - n_i\}.$$

Lemma 1 then yields the desired formula for H^*. □

The derivation of H^* is more complex in the present situation than for the special cases of Sections III.B and IV.B, partly because the group-theoretic lemma on pages 21-22 is no longer applicable. Indeed, when $S \neq S'$, it is not true that $d^*(S\pi,S'\sigma) = \min_{\beta \varepsilon S'\sigma} d(\pi,\beta)$, even if d is bi-invariant. However, there _is_ a somewhat analogous result, which will be helpful in the derivation of C^*.

Lemma 5: Let G be any finite group, d a metric on G, and $K \subset K'$ any subgroups of G. If d is left-invariant, then

$$\max_{\alpha \varepsilon K\pi} \min_{\beta \varepsilon K'\sigma} d(\alpha,\beta) \leq \max_{\beta \varepsilon K'\sigma} \min_{\alpha \varepsilon K\pi} d(\alpha,\beta) \quad \forall \pi,\sigma \varepsilon G.$$

Thus, for the case $G = S_n$, $K = S \subset K' = S'$, we have

$$d^*(S\pi,S'\sigma) = d(\pi_{min}(\sigma_{max}),\sigma_{max}).$$

Proof of Lemma 5: Let $m = \max_{\beta\varepsilon K'\sigma} \min_{\alpha\varepsilon K\pi} d(\alpha,\beta)$. Then

$$d(k_1\pi,\sigma) \leq m \quad \text{for some } k_1\varepsilon K$$

$$\Longrightarrow d(k_2\pi,k_2k_1^{-1}\sigma) \leq m \quad \forall k_2\varepsilon K$$

$$\Longrightarrow \forall k_2\varepsilon K\ \exists k' = k_2k_1^{-1} \varepsilon K\subset K'$$

$$\text{such that } d(k_2\pi,k'\sigma) \leq m$$

$$\Longrightarrow \max_{\alpha\varepsilon K\pi} \min_{\beta\varepsilon K'\sigma} d(\alpha,\beta) \leq m$$

as was to be shown. For the case $G = S_n$, $K = S\subset K' = S'$, the formula

$$d^*(S\pi,S'\sigma) = d(\pi_{min}(\sigma_{max}),\sigma_{max})$$

follows directly from the proof of Lemma 1. □

Proof of the Theorem for C^*: Recall that for C^*, the theorem only considers the case $S = S_{n-k}$, $S' = S_{n-k'}$, $k \geq k'$. By Lemma 5,

$$C^*(S_{n-k}\pi,\ S_{n-k'}\sigma) = C(\pi_{min}(\sigma_{max}),\sigma_{max}).$$

Also, by the right-invariance of C^*, it suffices to prove the theorem when π = id, i.e., to show

$$C^*(S_{n-k},\ S_{n-k'}\sigma) = k - \left[\begin{array}{l}\text{number of cycles in } \sigma\\ \text{consisting entirely of}\\ \text{numbers} \leq k'\end{array}\right]$$

when $n > k \geq k'$, and

$$C^*(S_{n-k}, S_{n-k'}\sigma) = (n-1) - \begin{bmatrix}\text{number of cycles in } \sigma \\ \text{consisting entirely} \\ \text{of numbers} \leq k'\end{bmatrix}$$

when $n = k > k'$. For any $\sigma_0 \varepsilon S'\sigma$, define a cycle of σ_0 to be of type 1 if all numbers in the cycle are $\leq k'$, of type 2 if all are $\leq k$ but at least one is $> k'$, and of type 3 if some number in the cycle is $> k$.

The first step in the derivation is the observation that

$$C(\pi_{min}(\sigma_0),\sigma_0) = k - \left(\begin{matrix}\text{number of cycles in} \\ \sigma_0 \text{ of types 1 and 2}\end{matrix}\right).$$

Indeed,

$$C(\pi_{min}(\sigma_0),\sigma_0) = \min_{\pi_0 \varepsilon S_{n-k}} C(\pi_0,\sigma_0) = \min_{\pi_0 \varepsilon S_{n-k}} C(\sigma_0,\pi_0)$$

$$= C^*(S_{n-k}\sigma_0, S_{n-k})$$

by the Lemma of Section III.B, pages 21-22, and

$$C^*(S_{n-k}\sigma_0, S_{n-k}) = C^*(S_{n-k}, S_{n-k}\sigma_0)$$

$$= k - \begin{bmatrix}\text{number of cycles in } \sigma_0 \\ \text{consisting entirely of} \\ \text{numbers} \leq k\end{bmatrix},$$

by the derivation of C^* in Section III.B for the case $S = S' = S_{n-k}$.

Thus

$$C^*(S_{n-k}, S_{n-k'}\sigma) = C(\pi_{min}(\sigma_{max}),\sigma_{max})$$

$$= \max_{\sigma_0 \varepsilon S_{n-k'}\sigma} C(\pi_{min}(\sigma_0),\sigma_0)$$

$$= k - \min_{\sigma_0 \varepsilon S_{n-k'}\sigma} \binom{\text{number of cycles in}}{\sigma_0 \text{ of types 1 and 2}}.$$

Now for any $\sigma_0 \varepsilon S_{n-k'}\sigma$, $\sigma_0^{-1}(i) = \sigma^{-1}(i)$ $\forall i \leq k'$, and thus any type 1 cycle of σ_0 is also a type 1 cycle of σ, and vice versa. Hence

$$C^*(S_{n-k}, S_{n-k'}\sigma) = k - \binom{\text{number of type 1}}{\text{cycles in } \sigma} - m,$$

where

$$m = \min_{\sigma_0 \varepsilon S_{n-k'}\sigma} (\text{number of type 2 cycles in } \sigma_0).$$

It remains to show $m = 0$ if $n > k \geq k'$, while $m = 1$ if $n = k > k'$.

Let

$$(d_1, \ldots, e_1), (d_2, \ldots, e_2), \ldots, (d_p, \ldots, e_p)$$

be an enumeration of the cycles of σ of types 2 and 3, where $d_i > k'$ for all $i = 1, \ldots, p$. If $n > k \geq k'$, define $\sigma_1 \varepsilon S_n$ to have the same type 1 cycles as σ, and to have one additional cycle, namely

$$(d_1, \ldots, e_1, d_2, \ldots, e_2, \ldots, d_p, \ldots, e_p).$$

Then $\sigma_1 \varepsilon S_{n-k'}\sigma$ (since $\sigma_1^{-1}(i) = \sigma^{-1}(i)$ $\forall i \leq k'$), and this last cycle is of type 3, so $m = 0$.

If $n = k > k'$, define $\sigma_1 \varepsilon S_n$ to have the same type 1 cycles as σ, and to have the one additional cycle

$$(d_1,\ldots,e_1,d_2,\ldots,e_2,\ldots,d_p,\ldots,e_p).$$

This is a type 2 cycle, and $\sigma_1 \varepsilon S_{n-k'}\sigma$ as before, so $m \leq 1$. On the other hand, since $n = k > k'$, any permutation must have at least one type 2 cycle, so $m = 1$.

This completes the proof of all parts of the main theorem. □

*   *   *

The results for F^*, R^*, and U^*, given in Section B for the special case $S = S'$, are immediate consequences of the more general versions derived here. To derive the result of Section B for T^*, one need only show that if $n_i = n_i'$ for all i, then

$$\sum_{\substack{i<i' \\ j\geq j'}} n_{ij}n_{i'j'} = \sum_{\substack{i\leq i' \\ j>j'}} n_{ij}n_{i'j'}.$$

Subtracting $\sum_{\substack{i<i' \\ j>j'}} n_{ij}n_{i'j'}$ from both sides, one must show

$$\sum_{\substack{i<i' \\ j}} n_{ij}n_{i'j} = \sum_{\substack{i \\ j>j'}} n_{ij}n_{ij'},$$

but this follows from the identity

$$\sum_j \left(\sum_i n_{ij}^2 + 2\sum_{i<i'} n_{ij}n_{i'j}\right) = \sum_j \left(\sum_i n_{ij}\right)^2 = \sum_j (n_j')^2 = \sum_i n_i^2$$

$$= \sum_i \left(\sum_j n_{ij}\right)^2 = \sum_i \left(\sum_j n_{ij}^2 + 2\sum_{j>j'} n_{ij}n_{ij'}\right).$$

The formula for H^* given in Section B is also a consequence of the present theorem, since $\nu_{ij} = \delta_{ij}n_j' = \delta_{ij}n_i$ when $S = S'$, and hence

$$H^*(S\pi,S\sigma)$$

$$= n\text{-min}\{\sum_{i=1}^{r}\sum_{j=1}^{r'} \max\{0, n_{ij}+\delta_{ij}n_j'-n_j'\},$$

$$\sum_{i=1}^{r}\sum_{j=1}^{r'} \max\{0, n_{ij}+\delta_{ij}n_i-n_i\}\}$$

$$= n\text{-min}\{\sum_{j=1}^{r'} n_{jj}, \sum_{i=1}^{r} n_{ii}\} = n - \sum_{i=1}^{r} n_{ii}.$$

The version of C^* derived in Section B is not a special case of the situation considered here.

It takes a bit more work to verify that the Hausdorff metrics of Section III.B, for the case $S = S' = S_{n-k}$, also follow from the results of the present section. First of all, note that for this case, $r = r' = k + 1$, $n_1 = \ldots = n_k = n_1' = \ldots = n_k' = 1$, and

$$n_{ij} = \#(\pi^{-1}(N_i)\cap\sigma^{-1}(N_j)) = \begin{cases} \delta_{\pi^{-1}(i),\sigma^{-1}(j)} & \text{if } i \le k \text{ and } j \le k \\ \delta_{\pi^{-1}(i)\varepsilon B} & \text{if } i \le k \text{ and } j = k+1 \\ \delta_{\sigma^{-1}(j)\varepsilon D} & \text{if } i = k+1 \text{ and } j \le k \\ \#E = n-k-h & \text{if } i = j = k+1 \end{cases}$$

where A, B, D, E, and h are defined in III.B. Hence, by the preceding paragraph,

$$H^*(S_{n-k}\pi, S_{n-k}\sigma) = n - \sum_{i=1}^{r} n_{ii} = n - n_{k+1,k+1} - \sum_{i=1}^{k} \delta_{\pi^{-1}(i),\sigma^{-1}(i)}$$

$$= n - (n-k-h) - \#\{i=1,\ldots,k: \pi^{-1}(i) = \sigma^{-1}(i)\}$$

$$= \#\{i=1,\ldots,k: \pi^{-1}(i) \neq \sigma^{-1}(i)\} + h,$$

which is the formula of Section III.B.

The version of $C^*(S_{n-k}\pi, S_{n-k}\sigma)$ described in Section III.B is an immediate corollary of the present section's theorem.

To verify the results of Section III.B for R^*, F^*, T^*, and U^*, it is easiest to return to Lemmas 3 and 4 of the current section, and to note that in the case at hand:

$$\sigma_{max} = \langle\sigma^{-1}(1),\ldots,\sigma^{-1}(k),N_{k+1,k+1},\pi^{-1}(p_h),\ldots,\pi^{-1}(p_1)\rangle,$$

$$\pi_{min}(\sigma_{max}) = \langle\pi^{-1}(1),\ldots,\pi^{-1}(k),\sigma^{-1}(s_1),\ldots,\sigma^{-1}(s_h),N_{k+1,k+1}\rangle,$$

$$\pi_{max} = \langle\pi^{-1}(1),\ldots,\pi^{-1}(k),N_{k+1,k+1},\sigma^{-1}(s_h),\ldots,\sigma^{-1}(s_1)\rangle,$$

$$\sigma_{min}(\pi_{max}) = \langle\sigma^{-1}(1),\ldots,\sigma^{-1}(k),\pi^{-1}(p_1),\ldots,\pi^{-1}(p_h),N_{k+1,k+1}\rangle,$$

where, as in III.B, $p_1 < \ldots < p_h$ is an ordering of the set $\bigcup_{i\in B}\{\pi(i)\}$ and $s_1 < \ldots < s_h$ is an ordering of $\bigcup_{i\in D}\{\sigma(i)\}$.

Note

$$A = \{\pi^{-1}(1),\ldots,\pi^{-1}(k)\}\cap\{\sigma^{-1}(1),\ldots,\sigma^{-1}(k)\},$$

$$B = \{\pi^{-1}(p_1),\ldots,\pi^{-1}(p_h)\},$$

$$D = \{\sigma^{-1}(s_1),\ldots,\sigma^{-1}(s_h)\},$$

while $N_{k+1,k+1}$ is an enumeration of the elements of E, in some fixed order.

By Lemmas 1, 3, and 4,

$$d^*(S_{n-k}\pi,\ S_{n-k}\sigma) = \max\{d(\pi_{min}(\sigma_{max}),\sigma_{max}),\ d(\sigma_{min}(\pi_{max}),\pi_{max})\}$$

for d = R, F, T, or U. The contribution to $R^2(\pi_{min}(\sigma_{max}),\sigma_{max})$ from A is $\sum_{i\varepsilon A}(\pi(i)-\sigma(i))^2$, from B is $\sum_{j=1}^{h}(n+1-j-p_j)^2$, from D is $\sum_{j=1}^{h}(k+j-s_j)^2$, and from E is $(n-k-h)h^2$. Thus

$$R^2(\pi_{min}(\sigma_{max}),\sigma_{max}) = \sum_{i\varepsilon A}(\pi(i)-\sigma(i))^2 + \sum_{j=1}^{h}(n+1-j-p_j)^2$$

$$+ \sum_{j=1}^{h}(k+j-s_j)^2 + (n-k-h)h^2.$$

Likewise, $R^2(\sigma_{min}(\pi_{max}),\pi_{max})$ is a sum of contributions: $\sum_{i\varepsilon A}(\pi(i)-\sigma(i))^2$ from A, $\sum_{j=1}^{h}(k+j-p_j)^2$ from B, $\sum_{j=1}^{h}(n+1-j-s_j)^2$ from D, and $(n-k-h)h^2$ from E. The formula of III.B for $R^*(S_{n-k}\pi,\ S_{n-k}\sigma)$ now follows.

The derivation of $F^*(S_{n-k}\pi,\ S_{n-k}\sigma)$ takes a similar path; the only change is that the squared terms above are replaced by absolute values, and hence some simplification of the terms involving p_j and s_j is possible, e.g.,

$$\sum_{j=1}^{h}|n+1-j-p_j| + \sum_{j=1}^{h}|k+j-s_j| = \sum_{j=1}^{h}(n+1-j-p_j) + (k+j-s_j)$$

$$= \sum_{j=1}^{h}\left(\frac{n+k+1}{2}-p_j\right) + \sum_{j=1}^{h}\left(\frac{n+k+1}{2}-s_j\right)$$

$$= \sum_{i\varepsilon B}\left(\frac{n+k+1}{2}-\pi(i)\right) + \sum_{i\varepsilon D}\left(\frac{n+k+1}{2}-\sigma(i)\right).$$

To check the formula in III.B for $T^*(S_{n-k}\pi, S_{n-k}\sigma)$, recall that $T(\alpha,\beta)$ is the number of ordered pairs (i,j) such that $\alpha(i) < \alpha(j)$ and $\beta(i) > \beta(j)$. The contribution to $T(\pi_{min}(\sigma_{max}),\sigma_{max})$:

from AxA is the number of pairs $(i,j)\varepsilon AxA$ such that $\pi(i) < \pi(j)$ and $\sigma(i) > \sigma(j)$,

from AxB is $\sum_{j=1}^{h} (k-p_j-(j-1)) = h(2k-h-1)/2 - \sum_{i\varepsilon B} \pi(i)$,

from AxD is $\sum_{j=1}^{r} (k-s_j-(j-1)) = h(2k-h-1)/2 - \sum_{i\varepsilon D} \sigma(i)$,

from BxB is $(\#B)\cdot(\#B-1) = h(h-1)$,

from BxD is $(\#B)\cdot(\#D) = h^2$,

from BxE is $(\#B)\cdot(\#E) = h(n-k-h)$,

and is zero from every other cross product of A, B, D, and E. Thus,

$$T(\pi_{min}(\sigma_{max}),\sigma_{max}) = \begin{bmatrix}\text{number of pairs } (i,j)\ AxA \\ \text{such that } \pi(i) < \pi(j) \text{ and} \\ \sigma(i) > \sigma(j)\end{bmatrix}$$

$$+ h(n+k - \frac{h-1}{2}) - \sum_{i\varepsilon B} \pi(i) - \sum_{i\varepsilon D} \sigma(i).$$

Similarly, one verifies that $T(\sigma_{min}(\pi_{max}),\pi_{max})$ is exactly the same sum, and the desired result follows.

Finally, the formula of Section III.B for $U^*(S_{n-k}\pi, S_{n-k}\sigma)$ can be verified by checking that the permutations α and β constructed there are $(\pi_{min}(\sigma_{max}))^{-1}\sigma_{max}$ and $(\sigma_{min}(\pi_{max}))^{-1}\pi_{max}$, respectively. Hence

$$U^*(S_{n-k}\pi, S_{n-k}\sigma) = \max\ \{ U(\alpha,id), U(\beta,id)\}$$

by Lemmas 1 and 4 and the right-invariance of U.

E. The Tied Ranks Approach to Metrizing Partially Ranked Data

1. A Description of the Tied Ranks Approach

This section explores a "common sense" approach to metrizing partially ranked data, based on "tied ranks." In any partial ranking, there are certain items which are "tied," in the sense that the ranker does not state any preferences among them. The "tied ranks approach" assigns to these tied items the <u>average</u> of the ranks which they would possess if they were distinguishable. By using these average ranks in place of ordinary ranks, we can sometimes show that a metric on full rankings induces a metric on partial rankings.

This approach is, of course, not new. See Kendall [K3], for instance, for a discussion of the tied ranks approach applied to Spearman's rho.

As an example, the tied ranks version H_t of Hamming distance will now be calculated. Suppose the given partial rankings are

$$\pi^* = \langle 1,3,(2,6),7,(4,5)\rangle$$

and

$$\sigma^* = \langle 1,4,(2,3),7,(5,6)\rangle,$$

where (as in III.B) the asterisks on π^* and σ^* indicate that these are <u>partial</u> rankings, and where the rounded parentheses denote ties, e.g., items 2 and 6 are tied for third and fourth places in the partial ranking π^*. We think of π^* as assigning rank 3.5 to items 2 and 6 and rank 6.5 to items 4 and 5, and similarly for σ^*, so that π^* and σ^* are the functions

$$\pi^*(1) = 1, \pi^*(3) = 2, \pi^*(2) = \pi^*(6) = 3.5, \pi^*(7) = 5, \pi^*(4) = \pi^*(5) = 6.5$$

and

$$\sigma^*(1) = 1, \sigma^*(4) = 2, \sigma^*(2) = \sigma^*(3) = 3.5, \sigma^*(7) = 5, \sigma^*(5) = \sigma^*(6) = 6.5.$$

The tied ranks Hamming distance between the partial rankings is then

$$H_t(\pi^*,\sigma^*) = \#\{i = 1,\ldots,n: \pi^*(i) \neq \sigma^*(i)\},$$

which equals 3 in this example.

Thus, the computation of H_t coincides with the usual calculation of the Hamming distance between two full permutations, except that tied ranks are used in place of ordinary ranks.

2. Relations among the Tied Ranks, Hausdorff, and Fixed Vector Metrics

In the example of the preceding section, the partial rankings π^* and σ^* can also be thought of as right cosets $S\pi$ and $S\sigma$ of the subgroup $S = S_1 \times S_1 \times S_2 \times S_1 \times S_2 \subset S_7$. Thus, besides computing the tied ranks Hamming distance H_t, we could also calculate the Hausdorff distance $H^*(S\pi,S\sigma)$ between the two partial rankings. It is easy to check that the Hausdorff distance is also equal to 3.

The following result demonstrates that this is more than just a rare coincidence. In fact, if we restrict our attention to partial rankings which are both "of the same type," there are several relations among the tied ranks, Hausdorff, and fixed vector metrics.

Proposition: Let H_t, C_t, F_t, and R_t be the tied ranks versions of the metrics H, C, F, and R. Then, for the problem of measuring the distance between two partial rankings of the <u>same</u> type:

(i) H_t coincides with the Hausdorff metric H^*.

(ii) C_t coincides with the Hausdorff metric C^*.

(iii) F_t coincides with the fixed vector metric F_{fv}.

(iv) R_t coincides with the fixed vector metric R_{fv}.

<u>Proof</u>: Given two partial rankings of type $n_1,\ldots,n_r$, let π^* and σ^* be the corresponding functions, as discussed in Section 1. Also, let $S\pi$ and $S\sigma$ be the cosets associated with these partial rankings, where $\pi,\sigma\in S_n$ and $S = S_{n_1} \times S_{n_2} \times \ldots \times S_{n_r}$. For $i = 1,\ldots,r$, let c_i be the arithmetic mean of the integers in the set N_i (defined on page 34). Then $c_1,\ldots,c_r$ are the values taken on by the functions π^* and σ^*. Moreover $\pi^{-1}(N_i) = \pi^{*-1}(c_i)$ and $\sigma^{-1}(N_i) = \sigma^{*-1}(c_i)$ for all $i = 1,\ldots,r$.

<u>Proof of (i)</u>: The tied ranks Hamming distance

$$
\begin{aligned}
H_t(\pi^*,\sigma^*) &= \#\{i=1,\ldots,n\colon \pi^*(i) \neq \sigma^*(i)\} \\
&= n - \#\{i=1,\ldots,n\colon \pi^*(i) = \sigma^*(i)\} \\
&= n - \sum_{i=1}^{r} \#[\pi^{*-1}(c_i) \cap \sigma^{*-1}(c_i)] \\
&= n - \sum_{i=1}^{r} \#[\pi^{-1}(N_i) \cap \sigma^{-1}(N_i)].
\end{aligned}
$$

This agrees with the Hausdorff distance $H^*(S\pi,S\sigma)$ derived in Section IV.B. □

<u>Proof of (ii)</u>: The tied ranks Cayley distance $C_t(\pi^*,\sigma^*)$ is the minimum number of transpositions of items necessary to transform π^* into σ^*.

Formally, this is the smallest number γ such that one can find transpositions $t_1,\ldots,t_\gamma$ satisfying $\pi^* t_1\ldots t_\gamma = \sigma^*$. On the other hand, the Hausdorff distance $C^*(S\pi,S\sigma) = \min_{\beta\varepsilon S\sigma} C(\pi,\beta)$ by the Lemma of Section III.B, pages 21-22. This is the smallest number γ' such that $\pi\tau_1\ldots\tau_{\gamma'}\varepsilon S\sigma$ for some transpositions $\tau_1,\ldots,\tau_{\gamma'}$.

To show that $C^* \leq C_t$, let $t_1,\ldots,t_\gamma$ be transpositions satisfying $\pi^* t = \sigma^*$, where $t = t_1\ldots t_\gamma$. Then for all $i = 1,\ldots,r$,

$$
\begin{aligned}
&(\pi^* t)^{-1}(c_i) = \sigma^{*-1}(c_i) \\
&\Longrightarrow t^{-1}\pi^{*-1}(c_i) = \sigma^{*-1}(c_i) \\
&\Longrightarrow t^{-1}\pi^{-1}(N_i) = \sigma^{-1}(N_i) \\
&\Longrightarrow (\pi t)^{-1}(N_i) = \sigma^{-1}(N_i) \\
&\Longrightarrow \pi t \varepsilon S\sigma.
\end{aligned}
$$

Hence $\gamma' \leq \gamma$, so $C^* \leq C_t$.

To prove that $C_t \leq C^*$, let $\tau_1,\ldots,\tau_{\gamma'}$ be transpositions satisfying $\pi\tau\varepsilon S\sigma$, where $\tau = \tau_1\ldots\tau_{\gamma'}$. Let $\beta = \pi\tau\varepsilon S\sigma$; then $\beta^{-1}(N_i) = \sigma^{-1}(N_i)$ for all $i = 1,\ldots,r$. Thus, $\pi\tau = \beta$ implies that for all $i = 1,\ldots,r$,

$$
\begin{aligned}
&(\pi\tau)^{-1}(N_i) = \beta^{-1}(N_i) = \sigma^{-1}(N_i) \\
&\Longrightarrow \tau^{-1}\pi^{-1}(N_i) = \sigma^{-1}(N_i) \\
&\Longrightarrow \tau^{-1}\pi^{*-1}(c_i) = \sigma^{*-1}(c_i) \\
&\Longrightarrow (\pi^*\tau)^{-1}(c_i) = \sigma^{*-1}(c_i) \\
&\Longrightarrow \pi^*\tau = \sigma^*.
\end{aligned}
$$

Hence $\gamma \leq \gamma'$, so $C_t \leq C^*$. $\square$

Proof of (iii) and (iv): This is immediate from the tied ranks interpretation of the fixed vector metrics F_{fv} and R_{fv}, as already discussed in Sections III.C and IV.C. □

3. Limitations of the Tied Ranks Approach

The tied ranks approach does not always produce a natural extension of a metric on full rankings to a metric on partial rankings. The purpose of this section is to demonstrate this fact, and to discuss certain advantages of the Hausdorff and fixed vector metrics.

One shortcoming of the tied ranks approach is that it sometimes does not yield a well-defined metric. As an example, consider Ulam's metric on full rankings:

$$U(\pi,\sigma) = n - \binom{\text{length of the longest increasing}}{\text{subsequence in } \sigma\pi^{-1}(1),\ldots,\sigma\pi^{-1}(n)}.$$

The tied ranks version of this metric would be

$$U_t(\pi^*,\sigma^*) = n - \binom{\text{length of the longest increasing}}{\text{subsequence in } \sigma^*\pi^{*-1}(1),\ldots,\sigma^*\pi^{*-1}(n)}.$$

However, $\sigma^*\pi^{*-1}$ is not well-defined because π^* is not a one-to-one function.

A second limitation of the tied ranks approach is that even when it is well-defined, it sometimes produces nonsensical results. This can be especially true when we are trying to measure the distance between two partial rankings "of different types." For instance, consider the tied ranks Hamming distance between two partial rankings π^* and σ^* of eight items. Suppose π^* and σ^* are of types 3,5 and 5,3, respectively, so that in π^*, three items are tied for first and five are tied for last, while in σ^*, five items are tied for first and three are tied for last. The tied ranks Hamming distance always equals eight, regardless of what π^* and σ^* are! The ranges of the functions π^* and σ^* are disjoint (range $(\pi^*) = \{2,6\}$ and range $(\sigma^*) = \{3,7\}$), and thus

$$H_t(\pi^*,\sigma^*) = \#\{i=1,\ldots,8:\ \pi^*(i) \neq \sigma^*(i)\} = 8.$$

As an even more extreme example, the tied ranks Cayley distance between two partial rankings of different types is always infinite! This is because the tied ranks Cayley distance $C_t(\pi^*,\sigma^*)$ is the minimum number of transpositions (of items) necessary to transform π^* into σ^*. But any such transposition preserves the type of the partial ranking. Therefore, even if we perform an arbitrarily large number of transpositions on π^*, we cannot transform it into σ^*, if the two partial rankings are of different types. A similar problem occurs with the tied ranks version of Kendall's tau, defined as the minimum number of pairwise adjacent transpositions needed to transform π^* into σ^*.

Of course, none of these problems arises with the Hausdorff metrics. As demonstrated in previous sections, this is a general approach for extending any metric to partial rankings.

Finally, even when a tied ranks distance "makes sense," the Hausdorff and fixed vector metrics can be of independent interest and importance. The tied ranks versions of Spearman's footrule and Spearman's rho are reasonable metrics, and, as shown previously, they coincide with the fixed vector metrics F_{fv} and R_{fv}, respectively. However, the fixed vector approach has the extra advantage of providing a whole host of metrics on partial rankings: there are many possible choices of the fixed vector $\underset{\sim}{v}$, with the tied ranks metrics corresponding to just one of these possible choices. Similarly, the Hausdorff metrics F^* and R^* are of independent interest, because they are reasonable "distance measures," and yet are distinct from the tied ranks metrics. As discussed in Chapter VI, these new metrics can be useful for various data-analytic purposes.

CHAPTER V - DISTRIBUTIONAL PROPERTIES OF THE METRICS

A. Exact Distributions

Given two partial rankings π^* and σ^*, both of the same type, we might ask whether the two rankings are "significantly" correlated. That is, do the two rankings show more agreement than one might expect, under the assumption that they were generated, independently, from a uniform distribution on all possible partial rankings?

Of course, one way to answer this is to compute a p-value for π^* and σ^*, as follows: let d^* be any one of our metrics on partial rankings. Find the exact probability distribution of d^*, under the uniformity assumption described above. In particular, calculate the probability that d^* is less than or equal to the <u>observed</u> value $d^*(\pi^*,\sigma^*)$. This probability is the p-value for π^* and σ^*: the smaller it is, the stronger the evidence that π^* and σ^* are "significantly" correlated.

Thus, to compute the p-value, we must first find the probability distribution of d^* under the foregoing uniformity assumption. For partially ranked data of the type considered in Chapter III, where each judge lists his k favorite items out of n, exact probability distributions have indeed been computed for forty different pairs (n,k), for each of the eight metrics T^*, R^*, R_{fv}, F^*, F_{fv}, C^*, H^*, and U^*. Many of these distributions are displayed in Appendix D.

For the Hausdorff metric H^* on S_n/S_{n-k}, it is also possible to derive the exact distribution by combinatorial methods:

Notation: For any positive integers $a \geq b$, let $a_{(b)}$ denote the b-fold product $a\cdot(a-1)\cdot\ldots\cdot(a-b+1)$.

Proposition: Under the assumption that π^* and σ^* are chosen independently and at random from all possible partial rankings in S_n/S_{n-k}, the probability distribution of the Hausdorff distance $H^*(\pi^*,\sigma^*)$ is given by:

$$P(H^* = m) = \begin{cases} \dfrac{1}{n_{(k)}} \displaystyle\sum_{\substack{(i,j):\\ i+j\leq k\\ 2i+j=m}} \binom{k}{i}(n-k)_{(i)}\binom{k-i}{j} \sum_{q=0}^{j} (-1)^q \binom{j}{q}(i+j-q)_{(j-q)} & \text{for } m=0,1,\ldots,\min\{2k,n\} \\ 0 & \text{otherwise.} \end{cases}$$

Proof: Since H^* is right-invariant, it may be assumed that σ^* is the coset of S_{n-k} containing the identity permutation. Thus $H^*(\pi^*,\sigma^*) = H^*(S_{n-k}\pi, S_{n-k})$, where π is assumed to be uniformly distributed over all possible permutations.

Now

$$\begin{aligned} H^*(S_{n-k}\pi,\ S_{n-k}) &= \#\{i=1,\ldots,k:\ \pi(i) \neq i\} + \#\{i=1,\ldots,k:\ \pi(i) > k\} \\ &= 2\#\{i=1,\ldots,k:\ \pi(i) > k\} \\ &\quad + \#\{i=1,\ldots,k:\ \pi(i) \leq k \text{ and } \pi(i) \neq i\} \\ &= 2I+J, \end{aligned}$$

where I and J are the random variables

$$I = \#\{i=1,\ldots,k: \pi(i) > k\}$$

and

$$J = \#\{i=1,\ldots,k: \pi(i) \leq k \text{ and } \pi(i) \neq i\},$$

respectively. Therefore

$$P(H^* = m) = P(2I + J = m) = \sum_{\substack{(i,j):\\ i+j\leq k\\ 2i+j=m}} P(I = i \cap J=j) = \sum_{\substack{(i,j):\\ i+j\leq k\\ 2i+j=m}} P(I=i)\,P(J=j|I=i).$$

It remains to evaluate $P(I=i)$ and $P(J=j|I=i)$. Clearly, I has a hypergeometric distribution, and

$$P(I=i) = \frac{\binom{n-k}{i}\binom{k}{k-i}}{\binom{n}{k}}.$$

To find $P(J=j|I=i)$, let E_1, E_2, and E_3 be the events

$$E_1 = \{\pi(h) > k \quad \forall h=1,\ldots,i$$

$$\text{and } \pi(h) \leq k \quad \forall h=i+1,\ldots,k\},$$

$$E_2 = \{\pi(h) \neq h \quad \forall h=i+1,\ldots,i+j\},$$

and

$$E_3 = \{\pi(h) = h \quad \forall h=i+j+1,\ldots,k\}.$$

Since π is assumed to be uniformly distributed over S_n, it follows by symmetry that

$$P(J=j|I=i) = P(J=j|E_1) = \binom{k-i}{j} P(E_2 \cap E_3|E_1)$$

$$= \binom{k-i}{j} P(E_3|E_1)\, P(E_2|E_1 \cap E_3) = \binom{k-i}{j} \frac{1}{k_{(k-i-j)}} P(E_2|E_1 \cap E_3).$$

Finally, to evaluate $P(E_2|E_1 \cap E_3)$, for $p = 1,\ldots,j$ let A_p be the event $\{\pi(i+p) = i+p\}$. Then, by the well-known combinatorial "principle of inclusion and exclusion" [F2, p. 106],

$$P(E_2|E_1 \cap E_3) = P(A_1^c \cap \ldots \cap A_j^c\, |E_1 \cap E_3)$$

$$= 1 - P(A_1 \cup \ldots \cup A_j|E_1 \cap E_3)$$

$$= 1 - \sum_{q=1}^{j} (-1)^{q+1} \binom{j}{q} P(A_1 \cap \ldots \cap A_q|E_1 \cap E_3)$$

$$= 1 - \sum_{q=1}^{j} (-1)^{q+1} \binom{j}{q} \frac{(i+j-q)_{(j-q)}}{(i+j)_{(j)}}$$

$$= \sum_{q=0}^{j} (-1)^{q} \binom{j}{q} \frac{(i+j-q)_{(j-q)}}{(i+j)_{(j)}}\,.$$

Putting together all of the pieces and cancelling a few terms yields the desired result. □

<u>Remark</u>: When k is small, the preceding proposition is actually useful for calculating the distribution of H^* on S_n/S_{n-k}. In fact:

$$\text{On } S_n/S_{n-1},\ H^* = \begin{cases} 0 & \text{with probability } \frac{1}{n} \\ 2 & \text{with probability } \frac{n-1}{n} \end{cases}$$

$$\text{On } S_n/S_{n-2},\ H^* = \begin{cases} 0 \text{ with probability } \dfrac{1}{n(n-1)} \\[2ex] 2 \text{ with probability } \dfrac{2(n-2)+1}{n(n-1)} \\[2ex] 3 \text{ with probability } \dfrac{2(n-2)}{n(n-1)} \\[2ex] 4 \text{ with probability } \dfrac{(n-2)(n-3)}{n(n-1)} \end{cases}$$

$$\text{On } S_n/S_{n-3},\ H^* = \begin{cases} 0 \text{ with probability } \dfrac{1}{n(n-1)(n-2)} \\[2ex] 2 \text{ with probability } \dfrac{3(n-3)+3}{n(n-1)(n-2)} \\[2ex] 3 \text{ with probability } \dfrac{6(n-3)+2}{n(n-1)(n-2)} \\[2ex] 4 \text{ with probability } \dfrac{3(n-3)(n-4)+3(n-3)\cdot 3}{n(n-1)(n-2)} \\[2ex] 5 \text{ with probability } \dfrac{3(n-3)(n-4)\cdot 2}{n(n-1)(n-2)} \\[2ex] 6 \text{ with probability } \dfrac{(n-3)(n-4)(n-5)}{n(n-1)(n-2)} \end{cases}$$

It may be checked that these theoretically obtained results agree with the computer-generated results in Appendix D.

B. Asymptotic Distributions

When n is large, computing the exact distribution of d^* may be impractical, and even the exact distributional formula for H^* becomes unwieldy. However, there is an <u>asymptotic</u> result for three of the metrics: H^*, F_{fv}, and R_{fv}^2 are asymptotically normally distributed.[1]

The main purpose of this section is to prove that this asymptotic normality holds under quite general conditions. In particular, consider the metrics H^*, F_{fv}, and R_{fv}, as defined on the quotient space S_n/S. These yield distances between partial rankings of type $n_1,\ldots,n_r$, where $n_1,\ldots,n_r$ is a partition of n. As n tends to infinity, the partition $n_1,\ldots,n_r$ must vary, of course. Yet quite unrestrictive assumptions about the limiting behavior of the partition guarantee the asymptotic normality of the metrics.

In what follows, d^*, S, r, n_i, λ_i, M, λ_M and almost every other parameter are implicitly understood to depend on n, although this fact is suppressed from the notation.

<u>Definition</u>: The metric d^* on S_n/S is asymptotically normally distributed if

$$\lim_{n\to\infty} P\left\{ \frac{d^*(\pi^*,\sigma^*) - Ed^*(\pi^*,\sigma^*)}{[\mathrm{Var}\ d^*(\pi^*,\sigma^*)]^{\frac{1}{2}}} \leq x\right\} = \Phi(x)$$

for all real numbers x, where Φ is the standard normal cumulative

1. In this section, for brevity of notation, the quantity R_{fv}^2 will be referred to as a "metric." Strictly speaking, this is not quite accurate: R_{fv}^2 is the square of a metric, and does not itself satisfy the triangle inequality.

distribution function. Here the foregoing probability, as well as the expectation and variance of $d^*(\pi^*,\sigma^*)$, are computed under the assumption that π^* and σ^* are generated, independently, from a uniform distribution on all possible partial rankings of type $n_1,\ldots,n_r$.

<u>Definition</u>: If $n_1,\ldots,n_r$ is a partition of n, let $\lambda_i = \frac{n_i}{n}$. Furthermore, let M be the index such that

$$n_M = \max\{n_1,\ldots,n_r\} \text{ and } \lambda_M = \max\{\lambda_1,\ldots,\lambda_r\}.$$

Thus, for a partial ranking of type $n_1,\ldots,n_r$, λ_M is the "maximum proportion of items which are tied for some position."

We can now state the main result of this section.

<u>Theorem</u>: If $\limsup_{n\to\infty} \lambda_M < 1$ and $\liminf_{n\to\infty} \lambda_M > 0$, then the metrics H^*, F_{fv}, and R^2_{fv} on S_n/S are asymptotically normally distributed. Moreover, the exact means and variances of these metrics are as follows:

$$E(H^*) = n - \sum_{i=1}^{r} \frac{n_i^2}{n}$$

$$\mathrm{Var}(H^*) = \sum_{i=1}^{r} \frac{n_i^2(n-n_i)^2}{n^2(n-1)} + 2\sum_{i<k} \frac{n_i^2 n_k^2}{n^2(n-1)}$$

$$E(F_{fv}) = \sum_{i=1}^{r}\sum_{j=1}^{r} |c_i-c_j| \frac{n_i n_j}{n}$$

$$\mathrm{Var}(F_{fv}) = \frac{1}{n^2(n-1)} \sum_{i=1}^{r}\sum_{j=1}^{r}\sum_{k=1}^{r}\sum_{m=1}^{r} n_i n_j n_k n_m |c_i-c_j|$$

$$\cdot\,(|c_i-c_j|+|c_k-c_m|-2|c_k-c_j|)$$

$$E(R^2_{fv}) = \sum_{i=1}^{r}\sum_{j=1}^{r} |c_i - c_j|^2 \frac{n_i n_j}{n}$$

$$\mathrm{Var}(R^2_{fv}) = \frac{1}{n^2(n-1)} \sum_{i=1}^{r}\sum_{j=1}^{r}\sum_{k=1}^{r}\sum_{m=1}^{r} n_i n_j n_k n_m |c_i - c_j|^2$$

$$\cdot\ (|c_i - c_j|^2 + |c_k - c_m|^2 - 2|c_k - c_j|^2)$$

where $c_i = n_1 + \ldots + n_{i-1} + \frac{n_i + 1}{2}$ is the average of the n_i numbers in the set $N_i = \{n_1 + \ldots + n_{i-1} + 1, \ldots, n_1 + \ldots + n_{i-1} + n_i\}$.

Remark: The theorem says that the metrics are asymptotically normal if λ_M is bounded away from 0 and 1 as n approaches infinity. Therefore, when this condition holds and when n is large, the metrics are approximately normally distributed, with means and variances as stated in the theorem.

David Pickard has a colorful way of describing the foregoing condition on λ_M: "The categories do not slip into oblivion, nor do they coalesce completely."

Proof of the Theorem:

1. Derivation of Means and Variances: We first derive the means and variances of the metrics, and begin by introducing some notation. Recall that for two *fixed* partial rankings $\pi^* = S\pi$ and $\sigma^* = S\sigma$, $n_{ij} = \#[\pi^{-1}(N_i) \cap \sigma^{-1}(N_j)]$ denotes the number of items ranked in the i-th category by π^* and in the j-th category by σ^*. If, now, we regard π^* and σ^* as being selected *randomly* and uniformly from all possible partial rankings, then the quantity n_{ij} becomes a *random variable*, which we denote by N_{ij}.

N_{ij} clearly has a hypergeometric distribution. Indeed, we may regard the n_i items in the set $\pi^{-1}(N_i)$ as being selected without replacement from a pool of n_j "good" items $(\sigma^{-1}(N_j))$ and $n-n_j$ "bad" items $(\sigma^{-1}(N_j)^c)$, and $N_{ij} = \#[\pi^{-1}(N_i) \cap \sigma^{-1}(N_j)]$ counts the number of "good" items selected. Therefore:

$$E(N_{ij}) = \frac{n_i n_j}{n}$$

$$Var(N_{ij}) = \frac{n_i n_j (n-n_i)(n-n_j)}{n^2(n-1)} .$$

Moreover, by the results of Sections IV.B and IV.C, the random variables $H^*(\pi^*,\sigma^*)$, $F_{fv}(\pi^*,\sigma^*)$, and $R^2_{fv}(\pi^*,\sigma^*)$ can be expressed in terms of the N_{ij}:

$$H^* = n - \sum_{i=1}^{r} N_{ii}$$

$$F_{fv} = \sum_{i=1}^{r} \sum_{j=1}^{r} |c_i - c_j| N_{ij}$$

$$R^2_{fv} = \sum_{i=1}^{r} \sum_{j=1}^{r} |c_i - c_j|^2 N_{ij}.$$

The formulas for $E(H^*)$, $E(F_{fv})$, and $E(R^2_{fv})$ are now immediate.

To derive the variances of these metrics, it is necessary to evaluate $Cov(N_{ij}, N_{km})$. Notice that $N_{ij} = X_{ij1} + \ldots + X_{ijn_i}$, where the random variables $X_{ij\alpha}$ are defined by:

$$X_{ij\alpha} = \begin{cases} 1 & \text{if } \sigma\pi^{-1}(n_1+\ldots+n_{i-1}+\alpha) \varepsilon N_j \\ 0 & \text{otherwise.} \end{cases}$$

Now, if $i = k$ but $j \neq m$, then

$$E(X_{ij\alpha}X_{im\beta}) = \begin{cases} \dfrac{n_j n_m}{n(n-1)} & \text{if } \alpha \neq \beta \\ 0 & \text{if } \alpha = \beta \end{cases}$$

and thus

$$E(N_{ij}N_{im}) = E[\sum_{\alpha=1}^{n_i} \sum_{\beta=1}^{n_i} X_{ij\alpha} X_{im\beta}] = \frac{n_i(n_i-1)n_j n_m}{n(n-1)} .$$

On the other hand, if $i \neq k$ and $j = m$, then

$$E(X_{ij\alpha}X_{kj\beta}) = \frac{n_j(n_j-1)}{n(n-1)} \quad \text{for all } \alpha,\beta,$$

and thus

$$E(N_{ij}N_{kj}) = E[\sum_{\alpha=1}^{n_i} \sum_{\beta=1}^{n_k} X_{ij\alpha} X_{kj\beta}] = \frac{n_i n_k n_j(n_j-1)}{n(n-1)} .$$

Finally, if both $i \neq k$ and $j \neq m$, then

$$E(X_{ij\alpha}X_{km\beta}) = \frac{n_j n_m}{n(n-1)} \quad \text{for all } \alpha,\beta,$$

and thus

$$E(N_{ij}N_{km}) = \frac{n_i n_k n_j n_m}{n(n-1)} .$$

Putting all of this together, we obtain, after some simplification,

$$\mathrm{Cov}(N_{ij}, N_{km}) = E(N_{ij} N_{km}) - E(N_{ij})\, E(N_{km}) = \begin{cases} \dfrac{n_i n_j n_m (n_i - n)}{n^2(n-1)} & \text{if } i=k,\ j\neq m \\[2ex] \dfrac{n_i n_j n_k (n_j - n)}{n^2(n-1)} & \text{if } i\neq k,\ j=m \\[2ex] \dfrac{n_i n_j n_k n_m}{n^2(n-1)} & \text{if } i\neq k,\ j\neq m. \end{cases}$$

It is now quite straightforward to find the variances of the metrics. For H^* we have

$$\mathrm{Var}(H^*) = \mathrm{Var}\,(n - \sum_{i=1}^{r} N_{ii}) = \mathrm{Var}\,(\sum_{i=1}^{r} N_{ii})$$

$$= \sum_{i=1}^{r} \mathrm{Var}\,(N_{ii}) + 2 \sum_{i<k} \mathrm{Cov}\,(N_{ii}, N_{kk})$$

$$= \sum_{i=1}^{r} \frac{n_i^2(n-n_i)^2}{n^2(n-1)} + 2 \sum_{i<k} \frac{n_i^2 n_k^2}{n^2(n-1)}$$

as was to be shown. Meanwhile,

$$\mathrm{Var}(F_{fv}) = \mathrm{Var}\,(\sum_{i=1}^{r} \sum_{j=1}^{r} |c_i - c_j| N_{ij})$$

$$= \sum_{ij} |c_i - c_j|^2\, \mathrm{Var}(N_{ij}) + \sum_{\substack{i \\ j\neq m}} |c_i - c_j||c_i - c_m| \mathrm{Cov}(N_{ij}, N_{im})$$

$$+ \sum_{\substack{j \\ i \neq k}} |c_i - c_j||c_k - c_j| \operatorname{Cov}(N_{ij}, N_{kj})$$

$$+ \sum_{\substack{i \neq k \\ j \neq m}} |c_i - c_j||c_k - c_m| \operatorname{Cov}(N_{ij}, N_{km})$$

$$= \sum_{ij} |c_i - c_j|^2 \frac{n_i n_j (n - n_i)(n - n_j)}{n^2(n-1)}$$

$$+ \sum_{\substack{i \\ j \neq m}} |c_i - c_j||c_i - c_m| \frac{n_i n_j n_m (n_i - n)}{n^2(n-1)}$$

$$+ \sum_{\substack{j \\ i \neq k}} |c_i - c_j||c_k - c_j| \frac{n_i n_j n_k (n_j - n)}{n^2(n-1)}$$

$$+ \sum_{\substack{i \neq k \\ j \neq m}} |c_i - c_j||c_k - c_m| \frac{n_i n_j n_k n_m}{n^2(n-1)} .$$

After some easy algebra, this reduces to the desired result:

$$\operatorname{Var}(F_{fv}) = \frac{1}{n^2(n-1)} \sum_{i=1}^{r} \sum_{j=1}^{r} \sum_{k=1}^{r} \sum_{m=1}^{r} n_i n_j n_k n_m |c_i - c_j|$$

$$\cdot \left(|c_i - c_j| + |c_k - c_m| - 2|c_k - c_j|\right).$$

The derivation of $\operatorname{Var}(R^2_{fv})$ is exactly the same as that of $\operatorname{Var}(F_{fv})$, except that terms of the form $|c_i - c_j|$ are replaced everywhere by $|c_i - c_j|^2$.

<u>2. Proof of the Asymptotic Normality Results</u>: First of all, by the right-invariance of the various metrics, the distribution of $d^*(\pi^*,\sigma^*)$ coincides with the distribution of $d^*((id)^*,\pi^*) = d^*(S,S\pi)$. Thus, it is enough to prove that $d^*(S,S\pi)$ is asymptotically normal, where $\pi\varepsilon S_n$ is uniformly distributed over all possible permutations.

The principal tool for proving asymptotic normality is Hoeffding's combinatorial central limit theorem [H3]. This states that if $(a_{IJ}^{(n)})$ is a sequence of n x n matrices, satisfying appropriate growth conditions, and if $\pi\varepsilon S_n$ is a random permutation, then the random variable $X_n = \sum_{I=1}^{n} a_{I\pi(I)}^{(n)}$ is asymptotically normally distributed. More precisely, the aforementioned "growth condition" that the sequence $(a_{IJ}^{(n)})$ must satisfy is:

$$\lim_{n\to\infty} \frac{\max_{IJ} (d_{IJ}^{(n)})^2}{\frac{1}{n} \sum_{I=1}^{n} \sum_{J=1}^{n} (d_{IJ}^{(n)})^2} = 0,$$

where $d_{IJ}^{(n)} = a_{IJ}^{(n)} - \frac{1}{n} \sum_{K=1}^{n} a_{KJ}^{(n)} - \frac{1}{n} \sum_{L=1}^{n} a_{IL}^{(n)} + \frac{1}{n^2} \sum_{K=1}^{n} \sum_{L=1}^{n} a_{KL}^{(n)}$.

By Theorem 2 of Hoeffding [H3, pp. 559-560],

$$\mathrm{Var}\,(X_n) = \frac{1}{n-1} \sum_{I=1}^{n} \sum_{J=1}^{n} (d_{IJ}^{(n)})^2,$$

so that the preceding condition is equivalent to:

$$(*) \qquad \lim_{n\to\infty} \frac{\max_{IJ} (d_{IJ}^{(n)})^2}{\mathrm{Var}\,(X_n)} = 0.$$

To apply Hoeffding's result to H^*, let the array $a_{IJ} = a_{IJ}^{(n)}$ be given by:

$$a_{IJ} = \begin{cases} 0 & \text{if there exists } i \text{ such that} \\ & I,J\varepsilon N_i = \{n_1+\ldots+n_{i-1}+1,\ldots,n_1+\ldots+n_{i-1}+n_i\} \\ 1 & \text{otherwise.} \end{cases}$$

Then $X_n = \sum_{I=1}^{n} a_{I\pi(I)}$ coincides with $H^*(S,S\pi)$. Moreover, a simple computation shows

$$d_{IJ} = \begin{cases} -1 + 2\dfrac{n_i}{n} - \displaystyle\sum_{i=1}^{r} \dfrac{n_i^2}{n^2} & \text{if } I,J \text{ are such that } I,J\varepsilon N_i \\ \\ \dfrac{n_i}{n} + \dfrac{n_j}{n} - \displaystyle\sum_{i=1}^{r} \dfrac{n_i^2}{n^2} & \text{if } I\varepsilon N_i, J\varepsilon N_j, \quad N_i \neq N_j; \end{cases}$$

hence $|d_{IJ}| \leq 2$ and $\max_{IJ}(d_{IJ})^2 \leq 4$. Finally,

$$\mathrm{Var}(X_n) = \mathrm{Var}(H^*) = \sum_{i=1}^{r} \frac{n_i^2(n-n_i)^2}{n^2(n-1)} + 2\sum_{i<k} \frac{n_i^2 n_k^2}{n^2(n-1)}$$

$$\geq \max_i \frac{n_i^2(n-n_i)^2}{n^2(n-1)} \geq \lambda_M^2\,(1-\lambda_M)^2\, n,$$

which is of order n, since λ_M is assumed to be bounded away from 0 and 1 as n approaches infinity. The condition (*) of Hoeffding's theorem is satisfied, so $X_n = H^*(S,S\pi)$ is asymptotically normally distributed.

To prove that F_{fv} is asymptotically normal, let the array a_{IJ} be given by:

$$a_{IJ} = |c_i - c_j|$$

where i and j are such that $I \varepsilon N_i$ and $J \varepsilon N_j$. Then

$$X_n = \sum_{I=1}^{n} a_{I\pi(I)}$$

coincides with $F_{fv}(S,S\pi)$. Furthermore, $\max_{IJ}(d_{IJ})^2$ is of order $\leq n^2$, since $a_{IJ} \leq n$ for all I and J.

Thus, to verify the condition (*) of Hoeffding's theorem, it will be sufficient to show that $\text{Var}(X_n) = \text{Var}(F_{fv})$ is of order $\geq n^3$. Letting $c_{ij} = |c_i - c_j|$, recalling our formula for $\text{Var}(F_{fv})$, and exploiting various symmetry properties, we have

$$\text{Var}(F_{fv}) = \frac{1}{n^2(n-1)} \sum_{i=1}^{r} \sum_{j=1}^{r} \sum_{k=1}^{r} \sum_{m=1}^{r} n_i n_j n_k n_m \, c_{ij}(c_{ij} + c_{km} - 2c_{kj})$$

$$= \frac{1}{n^2(n-1)} \sum_{i=1}^{r} \sum_{j=1}^{r} \sum_{k=1}^{r} \sum_{m=1}^{r} n_i n_j n_k n_m \, c_{ij}(c_{ij} + c_{km} - c_{kj} - c_{im})$$

$$= \frac{1}{2n^2(n-1)} \sum_{i=1}^{r} \sum_{j=1}^{r} \sum_{k=1}^{r} \sum_{m=1}^{r} n_i n_j n_k n_m (c_{ij} + c_{km})$$

$$\cdot \, [c_{ij} + c_{km} - (c_{kj} + c_{im})]$$

$$= \frac{1}{4n^2(n-1)} \sum_{i=1}^{r} \sum_{j=1}^{r} \sum_{k=1}^{r} \sum_{m=1}^{r} n_i n_j n_k n_m$$

$$\cdot \, [c_{ij} + c_{km} - (c_{kj} + c_{im})]^2.$$

Considering only those terms where i = m = M, the above yields

$$\mathrm{Var}(F_{fv}) \geq \frac{1}{4\, n^2(n-1)}\, n_M^2 \sum_{j=1}^{r} \sum_{k=1}^{r} n_j n_k \, (|c_M - c_j| + |c_M - c_k| - |c_j - c_k|)^2 .$$

It remains to show that the right hand side above is of order $\geq n^3$. Since λ_M is bounded away from 1, either $\sum_{j=1}^{M-1} n_j$ or $\sum_{j=M+1}^{r} n_j$ must be of order n; without loss of generality, assume the former is true. Then

$$\mathrm{Var}(F_{fv}) \geq \frac{1}{4n^2(n-1)}\, n_M^2 \sum_{j=1}^{M-1} \sum_{k=1}^{M-1} n_j n_k (|c_M - c_j| + |c_M - c_k| - |c_j - c_k|)^2$$

$$= \frac{1}{4n^2(n-1)}\, n_M^2 \sum_{j=1}^{M-1} \sum_{k=1}^{M-1} n_j n_k (c_M - c_j + c_M - c_k - |c_j - c_k|)^2$$

$$= \frac{1}{n^2(n-1)}\, n_M^2 \sum_{j=1}^{M-1} \sum_{k=1}^{M-1} n_j n_k (c_M - \max\{c_j, c_k\})^2 .$$

Recalling that c_i is the average value of the elements of N_i, we must have $c_M - \max\{c_j, c_k\} \geq \frac{n_M}{2}$, and therefore

$$\mathrm{Var}(F_{fv}) \geq \frac{n_M^4}{4n^2(n-1)} \left(\sum_{j=1}^{M-1} n_j\right)^2 .$$

Finally, since λ_M is bounded away from 0, $n_M^4 = (\lambda_M n)^4$ is of order n^4. This completes the proof that $\mathrm{Var}(F_{fv})$ is of order $\geq n^3$, and hence that F_{fv} is asymptotically normal.

The proof that R_{fv}^2 is asymptotically normal is exactly the same, except that terms of the form $|c_i - c_j|$ are replaced everywhere by $|c_i - c_j|^2$. To verify the condition (*) of Hoeffding's theorem, one checks that $\max_{IJ} (d_{IJ})^2$ is of order $\leq n^4$, while $\mathrm{Var}(R_{fv}^2)$ is of order $\geq n^5$, using arguments entirely similar to those for F_{fv}. $\square$

CHAPTER VI - DATA ANALYSIS, USING THE METRICS

Several applications of metrics on <u>fully</u> ranked data have already been proposed by various authors, especially Mallows [M1], Diaconis [D2], and Friedman and Rafsky [F3,F4]. These applications include fitting probability models to fully ranked data, clustering ranked data for a goodness-of-fit test, multidimensional scaling for fully ranked data, and testing for significant differences between two populations of rankers. A nice presentation of the uses of metrics on fully ranked data is in Diaconis' monograph [D2].

In the present chapter, these applications of metrics are extended to partially ranked data, and are illustrated by analyses of real data sets.

As a first step toward carrying out the data analyses, the metrics derived in Chapters III and IV were implemented in various FORTRAN subroutines. These are presented in Appendix B, together with a description of the underlying algorithms. An elegant $O(n^2)$ algorithm due to Peter H. Sellers [S1] was needed in order to program the metric U^* efficiently; for details see Appendix B.

A. Fitting Probability Models to Partially Ranked Data

1. Mallows' Model for Fully Ranked Data

At a symposium on ranking methods, Sir Maurice Kendall once stated that a major outstanding problem was to construct suitable non-null probability models for ranked data [K2]. Thus, Kendall wanted to model the situation where one has a large number of judges, each of whom ranks the same set of n items. By "non-null ranking models," Kendall meant models distinct from the uniform model, where all n! possible rankings of the n items are equally likely.

In response to this challenge, Colin Mallows proposed a practical non-null probability model for fully ranked data [M1], which utilizes the metrics of Section II.B. The model specifies a particular ranking $\pi_0 \varepsilon S_n$, which can be interpreted as "the most likely ranking of the n items," and further states that the probability of any other ranking π decreases exponentially according to the <u>distance</u> from π to π_0. More precisely, the model is:

$$\text{Prob}(\pi) = Ce^{-\lambda d(\pi,\pi_0)} \quad \forall \pi \varepsilon S_n,$$

where d is a <u>metric</u> on S_n, $\pi_0 \varepsilon S_n$ is the "location parameter" for this model, $\lambda \geq 0$ is a dispersion parameter, and

$$C = C(\lambda) = \Big(\sum_{\pi \varepsilon S_n} e^{-\lambda d(\pi,\pi_0)} \Big)^{-1}$$

is a constant of proportionality. The model is centered about the ranking π_0, and as λ increases the distribution becomes more and more peaked about π_0.

In his original Biometrika article, Mallows shows that this model arises in a natural way, from looking at paired comparisons. Indeed, suppose that the judge obtains a ranking of the n items as follows: he first makes all the $\binom{n}{2}$ possible comparisons between pairs of items, independently of each other, but then only accepts the results of these comparisons if they are consistent with a full ranking of the n items. Assume further that the pairwise comparisons are made according to the rule

$$(*) \qquad \text{Prob (item i preferred to item j)} = \begin{cases} p & \text{if } \pi_0(i) < \pi_0(j) \\ 1-p & \text{if } \pi_0(i) > \pi_0(j) \end{cases}$$

where $p \geq \frac{1}{2}$ is some fixed constant independent of i and j; the interpretation being that there is some fixed probability p that the judge's comparison of items i and j will result in the same preference as that given by the central ranking π_0. Then Mallows shows that the above procedure gives rise to the ranking model

$$\text{Prob}(\pi) = Ce^{-\lambda T(\pi,\pi_0)} \qquad \forall \pi \varepsilon S_n,$$

where $\lambda = \log\left(\frac{p}{1-p}\right) \geq 0$, and where T is the metric on S_n given by Kendall's tau. Of course, this is a special case of the more general Mallows model introduced earlier.

If, instead, the pairwise comparisons are made according to rules other than (*), Mallows obtains a ranking model of the form

$$\text{Prob}(\pi) = Ce^{-\lambda d(\pi,\pi_0)} \qquad \forall \pi \varepsilon S_n,$$

where d is a convex combination of Kendall's tau and Spearman's rho. Diaconis' monograph [D2] puts forth the idea of allowing d to be an arbitrary metric on S_n.

* * *

Paul Feigin and Ayala Cohen have successfully applied Mallows' model to a real ranked data set [F1]. A group of 143 Technion graduates was asked to rank ten occupations according to the degree of social prestige they associated with each. Feigin and Cohen fit the parameters π_0 and λ in Mallows' model by maximum likelihood, and for the metric d they used Kendall's tau. They concluded that Mallows' model provided a good fit to the data.

2. The Extension of Mallows' Model to Partially Ranked Data

Endowed with the metrics of Chapters III and IV, it is relatively straightforward to extend Mallows' model to _partial_ rankings of type $n_1,\ldots,n_r$. As in Chapter IV, let $S = S_{n_1} \times \ldots \times S_{n_r}$, so that each partial ranking corresponds to a member of the coset space S_n/S. The proposed generalization of Mallows' model is to require

$$P(\pi^*) = Ce^{-\lambda d^*(\pi^*,\pi_0^*)}$$

for all partial rankings $\pi^* \varepsilon S_n/S$. Here d^* is any of the derived metrics on S_n/S; that is, d^* can be either a Hausdorff metric or a fixed vector metric. As before, $\pi_0^* \varepsilon S_n/S$ is a "location parameter" for this model, and $\lambda \geq 0$ is a dispersion parameter. The model is centered about the partial ranking π_0^*, and as λ increases the distribution becomes more concentrated about π_0^*.

Remark on Notation: Previously, the coset notation $S\pi$ has been used to denote a partial ranking, as an aid to understanding the Hausdorff metrics, and to emphasize the group-theoretic approach. In this data analysis chapter, however, it seems more appropriate to denote the partial ranking by the simpler notation π^*, where it is implicitly understood that $\pi^* = S\pi\varepsilon S_n/S$ is actually a right coset.

3. A Likelihood Ratio Interpretation of the Triangle Inequality

It is natural to wonder why one should insist that d^* satisfy all the axioms for a metric. In particular, what is the interpretation of the triangle inequality? To answer this question, consider the following hypothesis testing situation: we wish to select a Mallows model, and the dispersion parameter λ is known and fixed, but there are two candidates π_A^* and π_B^* for the central partial ranking π_0^*. On the basis of a single observed partial ranking π^*, we wish to test the null hypothesis H_A: $\pi_0^* = \pi_A^*$ against the alternative hypothesis H_B: $\pi_0^* = \pi_B^*$. By the Neyman-Pearson Lemma, the most powerful test is to accept H_A for large values of the likelihood ratio statistic

$$\frac{P(\pi^*|\pi_A^*)}{P(\pi^*|\pi_B^*)} = \frac{Ce^{-\lambda d^*(\pi^*,\pi_A^*)}}{Ce^{-\lambda d^*(\pi^*,\pi_B^*)}}$$

$$= e^{-\lambda[d^*(\pi^*,\pi_A^*) - d^*(\pi^*,\pi_B^*)]}$$

It seems reasonable to insist that the most compelling evidence for H_A should occur when the <u>observed</u> partial ranking π^* is, in fact, the same as the <u>proposed</u> central ranking π_A^*. Thus, the likelihood ratio

$$e^{-\lambda[d^*(\pi^*,\pi_A^*) - d^*(\pi^*,\pi_B^*)]}$$

should be maximized when $\pi^* = \pi_A^*$; or, equivalently, we should have

$$d^*(\pi_A^*,\pi_A^*) - d^*(\pi_A^*,\pi_B^*) \leq d^*(\pi^*,\pi_A^*) - d^*(\pi^*,\pi_B^*)$$

for all partial rankings $\pi^* \varepsilon S_n/S$. Since $d^*(\pi_A^*,\pi_A^*) = 0$, the above condition amounts to the triangle inequality for d^*:

$$d^*(\pi^*,\pi_B^*) \leq d^*(\pi^*,\pi_A^*) + d^*(\pi_A^*,\pi_B^*).$$

4. Maximum Likelihood Estimation for the Model

Suppose we are given a data set consisting of partial rankings $\pi_1^*,\ldots,\pi_N^*$, all of the same type, and we wish to fit a Mallows model to this data by the method of maximum likelihood. As in Feigin and Cohen [F1, p. 209], the maximum likelihood estimate $\hat{\pi}_0^*$ is that partial ranking π^* which minimizes $\sum_{i=1}^{N} d^*(\pi_i^*,\pi^*)$. Meanwhile, the maximum likelihood estimate $\hat{\lambda}$ is that value of λ which makes the observed mean distance $\frac{1}{N}\sum_{i=1}^{N} d^*(\pi_i^*,\hat{\pi}_0^*)$ agree with the expected distance $E_{\hat{\pi}_0^*,\lambda}\, d^*(\cdot,\hat{\pi}_0^*)$. Here $E_{\hat{\pi}_0^*,\lambda}$ denotes taking expectation with respect to the Mallows model with parameters $\hat{\pi}_0^*$ and λ.

To derive these results, note that the likelihood function is

$$\prod_{i=1}^{N} [C(\lambda)e^{-\lambda d^*(\pi_i^*,\pi^*)}] = C(\lambda)^N e^{-\lambda \sum_{i=1}^{N} d^*(\pi_i^*,\pi^*)}.$$

For any fixed $\lambda \geq 0$, this is clearly maximized by choosing π^* so as to minimize $\sum_{i=1}^{N} d^*(\pi_i^*,\pi^*)$. Then, to maximize with respect to λ, one sets

$$\begin{aligned} 0 &= \frac{d}{d\lambda}[C(\lambda)^N e^{-\lambda \sum_{i=1}^{N} d^*(\pi_i^*,\hat{\pi}_0^*)}] \\ &= -[\sum_{i=1}^{N} d^*(\pi_i^*,\hat{\pi}_0^*)]C(\lambda)^N \cdot e^{-\lambda \sum_{i=1}^{N} d^*(\pi_i^*,\hat{\pi}_0^*)} \\ &\qquad + NC(\lambda)^{N-1}C'(\lambda) \cdot e^{-\lambda \sum_{i=1}^{N} d^*(\pi_i^*,\pi_0^*)}, \end{aligned}$$

yielding

$$(*) \qquad \frac{1}{N}\sum_{i=1}^{N} d^*(\pi_i^*,\hat{\pi}_0^*)\, C(\lambda) = C'(\lambda).$$

But

$$\sum_{\sigma^*\varepsilon S_n/S} C(\lambda)\, e^{-\lambda d^*(\sigma^*,\hat{\pi}_0^*)} = 1,$$

hence

$$C(\lambda) = \Big(\sum_{\sigma^*\varepsilon S_n/S} e^{-\lambda d^*(\sigma^*,\hat{\pi}_0^*)}\Big)^{-1}$$

and

$$C'(\lambda) = \sum_{\sigma^*\varepsilon S_n/S} d^*(\sigma^*,\hat{\pi}_0^*)\, e^{-\lambda d^*(\sigma^*,\hat{\pi}_0^*)}$$

$$\cdot\Big(\sum_{\sigma^*\varepsilon S_n/S} e^{-\lambda d^*(\sigma^*,\hat{\pi}_0^*)}\Big)^{-2}$$

$$= E_{\hat{\pi}_0^*,\lambda}\, d^*(\cdot,\hat{\pi}_0^*)\, C(\lambda).$$

Substituting back into (*) gives the condition

$$\frac{1}{N}\sum_{i=1}^{N} d^*(\pi_i^*,\hat{\pi}_0^*) = E_{\hat{\pi}_0^*,\lambda}\, d^*(\cdot,\hat{\pi}_0^*).$$

By the second derivative test, one can check that this does, indeed, give the maximizing value of λ.

In practice, the M.L.E.'s $\hat{\pi}_0^*$ and $\hat{\lambda}$ must be computed iteratively. To find $\hat{\pi}_0^*$, one makes an initial guess at this central partial ranking, and then modifies the guess by successive pairwise transpositions, until $\sum_{i=1}^{N} d^*(\pi_i^*, \cdot)$ is minimized. Then, to find $\hat{\lambda}$, the expected distance $E_{\hat{\pi}_0^*,\lambda}\, d^*(\cdot, \hat{\pi}_0^*)$ must be computed numerically for several values of λ, iterating until it agrees with the observed mean distance. FORTRAN subroutines for fitting these parameters are presented and discussed in Appendix C.

To conclude this section, it will be shown that the maximum likelihood estimate $\hat{\pi}_0^*$ satisfies an important "common sense" property. This result will be of interest in its own right, and will also be helpful in the next section, for deriving the asymptotic distribution of a goodness-of-fit statistic.

Suppose there exists a partial ranking π_b^* such that the observed data is concentrated about π_b^*, in the following sense: the closer a partial ranking is to π_b^*, the more often it occurs in the data set. Thus, in Figure 1, the circles represent contours of constant distance from π_b^*, and it is assumed that the partial rankings (represented by dots) on an inner circle have been observed more frequently than any of the partial rankings on a more exterior circle.

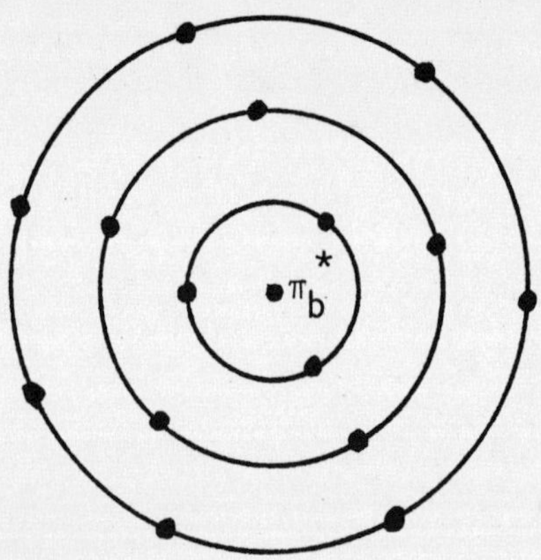

Figure 2. Pictorial Representation of π_b^*.

Under these circumstances, it is natural to expect that π_b^* should be the maximum likelihood estimate of π_0^*. The following result confirms this expectation.

<u>Notation</u>: For any partial ranking $\pi^* \varepsilon S_n/S$, let $N(\pi^*)$ denote the number of times that π^* has been observed in the data set $\pi_1^*, \ldots, \pi_N^*$.

<u>Proposition</u>: Suppose there exists $\pi_b^* \varepsilon S_n/S$, such that $N(\pi^*) \geq N(\sigma^*)$ whenever $d^*(\pi^*, \pi_b^*) < d^*(\sigma^*, \pi_b^*)$. Then $\pi_b^* = \hat{\pi}_0^*$.

<u>Proof</u>: The proof relies on the following simple observation.

<u>Lemma</u>: Let $a_1, \ldots, a_q$ and $b_1, \ldots, b_q$ be any numbers such that $a_1 \geq \ldots \geq a_q$ and $b_1 \leq \ldots \leq b_q$. Then

$$\sum_{i=1}^{q} a_i b_i \leq \sum_{i=1}^{q} a_i b_{\alpha(i)}$$

for any permutation $\alpha \varepsilon S_q$.

<u>Proof of Lemma</u>: The proof is by induction on q. The lemma is trivially true for q = 1. Now, assuming that it holds for q-1, it will be shown that it holds for q. For i = 1,...,q-1, define

$$\beta(i) = \begin{cases} \alpha(i) & \text{if } \alpha(i) < \alpha(q) \\ \\ \alpha(i)-1 & \text{if } \alpha(i) > \alpha(q). \end{cases}$$

Then $\beta \varepsilon S_{q-1}$, so by the inductive hypothesis, $\sum_{i=1}^{q-1} a_i b_{\beta(i)} \geq \sum_{i=1}^{q-1} a_i b_i$. Hence

$$\sum_{i=1}^{q} a_i b_{\alpha(i)} - \sum_{i=1}^{q} a_i b_i$$

$$= (\sum_{i=1}^{q} a_i b_{\alpha(i)} - \sum_{i=1}^{q-1} a_i b_{\beta(i)}) + (\sum_{i=1}^{q-1} a_i b_{\beta(i)} - \sum_{i=1}^{q-1} a_i b_i)$$

$$+ (\sum_{i=1}^{q-1} a_i b_i - \sum_{i=1}^{q} a_i b_i)$$

$$\geq (a_q b_{\alpha(q)} + \sum_{i:\alpha(i)>\alpha(q)} a_i (b_{\alpha(i)} - b_{\alpha(i)-1})) + 0 + (-a_q b_q)$$

$$\geq (a_q b_{\alpha(q)} + a_q \sum_{i:\alpha(i)>\alpha(q)} (b_{\alpha(i)} - b_{\alpha(i)-1})) - a_q b_q$$

$$= a_q b_{\alpha(q)} + a_q (b_q - b_{\alpha(q)}) - a_q b_q = 0. \square$$

Proof of the Proposition: Let q be the cardinality of S_n/S, and let $\sigma_1^*,\ldots,\sigma_q^*$ be an enumeration of S_n/S, such that $N(\sigma_i^*) \geq N(\sigma_{i+1}^*)$ and $d^*(\sigma_i^*,\pi_b^*) \leq d^*(\sigma_{i+1}^*,\pi_b^*)$ for all $i = 1,\ldots,q - 1$. To prove that π_b^* is the M.L.E. of π_0^*, it must be shown that π_b^* minimizes

$$\sum_{i=1}^{N} d^*(\pi_i^*,\cdot) = \sum_{i=1}^{q} N(\sigma_i^*)\, d^*(\sigma_i^*,\cdot).$$

Equivalently, for any $\pi_e^* \varepsilon S_n/S$, it must be shown that

$$\sum_{i=1}^{q} N(\sigma_i^*)\, d^*(\sigma_i^*,\pi_b^*) \leq \sum_{i=1}^{q} N(\sigma_i^*)\, d^*(\sigma_i^*,\pi_e^*).$$

For $i = 1,\ldots,q$, let $a_i = N(\sigma_i^*)$, $b_i = d^*(\sigma_i^*,\pi_b^*)$, and $e_i = d^*(\sigma_i^*,\pi_e^*)$. Then there exists a permutation $\alpha \varepsilon S_q$ such that

$e_i = b_{\alpha(i)}$ for all i, since

$$e_i = d^*(\sigma_i{}^*, \pi_e{}^*) = d^*(S\sigma_i, S\pi_e) = d^*(S\sigma_i\pi_e^{-1}\pi_b, S\pi_b)$$

$$= d^*(\ (\sigma_i\pi_e^{-1}\pi_b)^*,\ \pi_b{}^*)$$

by the right-invariance of d^*. Hence, by the lemma,

$$\sum_{i=1}^{q} a_i b_i \leq \sum_{i=1}^{q} a_i e_i ,$$

as was to be shown. □

5. A Goodness-of-Fit Result

In section 6, to measure how well the observed data fits a proposed Mallows model, a statistic of the form

$$X^2 = \sum_{i=1}^{M} \frac{(O_i - E_i)^2}{E_i}$$

will be used. The purpose of this section is to determine the asymptotic distribution of X^2, under the null hypothesis that the model is valid.

The statistic X^2 has, of course, been used to test goodness-of-fit for many different probability models. The index i ranges over M possible cells (or categories) into which the data may fall, O_i is the number of observations falling into the i-th cell, and E_i is the expected number of such observations under the proposed model. If a model has p continuous parameters which are estimated from the data by maximum likelihood, then under quite general conditions, it is well known that X^2 is asymptotically distributed as a chi-squared statistic with M-p-1 degrees of freedom [L1].

The extension of Mallows' model has one continuous parameter λ, so one would expect the asymptotic distribution of X^2 to have at most M-2 degrees of freedom. In addition to the one continuous parameter, there are several discrete parameters in π_0^*, which also must be estimated from the data. It is perhaps surprising that these additional discrete parameters do not affect the asymptotic distribution of X^2:

<u>Proposition</u>: Suppose the parameters π_0^* and λ in the extension of Mallows' model are estimated from the data by maximum likelihood. Suppose further that the null hypothesis holds; that is, the Mallows model

$$P(\pi^*) = Ce^{-\lambda d^*(\pi^*,\pi_0^*)} \qquad \forall \pi^* \varepsilon S_n/S$$

is valid for some $\pi_0^* \varepsilon S_n/S$ and some $\lambda > 0$. Then X^2 has, asymptotically, a chi-squared distribution with M-2 degrees of freedom.

<u>Proof</u>: Let Z be a χ^2_{M-2} random variable. To show that, as the sample size N tends to infinity, X^2 converges in distribution to a χ^2_{M-2} distribution, one must show that for any real number t,

$$\lim_{N\to\infty} P(X^2<t) = P(Z<t).$$

This will follow from Equations (i), (ii), (iii), and (iv) below.

First of all, notice that under the null hypothesis, one has $P(\pi^*) > P(\sigma^*)$ for any two partial rankings π^*,σ^* such that $d^*(\pi^*,\pi_0^*) < d^*(\sigma^*,\pi_0^*)$. Thus, by the Law of Large Numbers, $\lim_{N\to\infty} P(N(\pi^*) > N(\sigma^*)) = 1$ for any such π^* and σ^*. Here, as in Section 4, $N(\pi^*)$ denotes the number of times that π^* occurs in the sample $\pi_1^*,\ldots,\pi_N^*$. Hence, by the Proposition of Section 4, $\lim_{N\to\infty} P(\hat{\pi}_0^* = \pi_0^*) = 1$, and $\lim_{N\to\infty} P(\hat{\pi}_0^* \neq \pi_0^*) = 0$.

But

$$P(X^2<t) = P(X^2<t \mid \hat{\pi}_0^* = \pi_0^*)\, P(\hat{\pi}_0^* = \pi_0^*)$$

$$+ P(X^2<t \mid \hat{\pi}_0^* \neq \pi_0^*)\, P(\hat{\pi}_0^* \neq \pi_0^*),$$

and therefore:

$$\text{(i)} \quad \lim_{N\to\infty} \frac{P(X^2<t)}{P(X^2<t \mid \hat{\pi}_0^* = \pi_0^*)} = 1.$$

Next, define Y^2 to be the goodness-of-fit statistic which results when the true value of π_0^* is <u>known</u>, and only λ is fit by maximum likelihood. Thus,

$$Y^2 = \sum_{i=1}^{M} \frac{(O_i - E_i')^2}{E_i'},$$

where the E_i' are expected values with respect to the <u>known</u> parameter π_0^* and the estimated parameter λ. Notice $X^2 = Y^2$ whenever the data is such that $\hat{\pi}_0^* = \pi_0^*$, and hence

$$\text{(ii)} \quad P(X^2<t \mid \hat{\pi}_0^* = \pi_0^*) = P(Y^2<t \mid \hat{\pi}_0^* = \pi_0^*).$$

Moreover, since $\lim_{N\to\infty} P(\hat{\pi}_0^* = \pi_0^*) = 1$, one has

$$\text{(iii)} \quad \lim_{N\to\infty} \frac{P(Y^2<t)}{P(Y^2<t \mid \hat{\pi}_0^* = \pi_0^*)} = 1,$$

by the same argument which proved (i).

By equations (i), (ii), and (iii), the proof of the proposition can be finished by showing:

$$\text{(iv)} \quad \lim_{N\to\infty} P(Y^2<t) = P(Z<t).$$

Since only the continuous parameter λ must be estimated to calculate Y^2, the usual classical theory applies. By Theorem 5.1, page 148 of Lancaster [L1], the asymptotic distribution of Y^2 is χ^2_{M-2} if, for all $\pi^* \varepsilon S_n/S$, the expressions

$$\frac{d}{d\lambda} P(\pi^*) = [C'(\lambda) - d^*(\pi^*, \pi_0^*)\, C(\lambda)]\, e^{-\lambda d^*(\pi^*, \pi_0^*)}$$

and

$$\frac{d^2}{d\lambda^2} P(\pi^*) = [C''(\lambda) - 2d^*(\pi^*, \pi_0^*)\, C'(\lambda) + d^{*2}(\pi^*, \pi_0^*)\, C(\lambda)]\, e^{-\lambda d^*(\pi^*, \pi_0^*)}$$

are bounded in any finite interval. These conditions are easily verified, completing the proof of (iv) and the proposition. □

6. An Example: The Educational Testing Service Word Association Data

In this section, the extension of Mallows' model is fit to some ranked data provided by the Educational Testing Service [P1].

One hundred thirty black students were each asked to rank the five words "emotion," "author," "experimentation," "thought," and "vigil" according to their degree of association with the word "obsession." Upon examination of the data, it is observed that the students displayed a strong tendency to rank the words "emotion," "thought," and "author" as their first, second, and fifth choices, respectively; but the students were indifferent (evenly divided) as to whether "experimentation" or "vigil" should rank third. For modelling purposes, therefore, it seems appropriate to reduce the data to partial rankings of type 1,1,2,1, and to use one of the metrics on the corresponding coset space S_n/S, where

$$S = S_1 \times S_1 \times S_2 \times S_1 = \{\pi \varepsilon S_5 : \pi(i) = i \quad \forall\, i \neq 3,4\}.$$

That is, the first, second, and fifth choices of each student are recorded separately, but the distinction between third and fourth choices is ignored, so these are lumped together into a single category.

The extension of Mallows' model was fit to these partial rankings, taking the metric d^* to be the fixed vector version F_{fv} of Spearman's footrule. The results are shown in Table 1. The maximum likelihood estimate of the "location parameter" π_0^* is

$$\hat{\pi}_0^* = \begin{array}{l}\text{emotion (first choice),}\\ \text{thought (second choice),}\\ \text{author (fifth choice)}\end{array}$$

Table 1. Fitting a Mallows Model to the Educational Testing Service Data, Using the Metric F_{fv} on $S_5/S_1 \times S_1 \times S_2 \times S_1$.

Location Estimate $\hat{\pi}_0^*$: ETA

Dispersion Estimate $\hat{\lambda}$: 0.507

Partial Ranking	Distance to $\hat{\pi}_0^*$	Fitted Frequency	Observed Frequency	Partial Ranking	Distance to $\hat{\pi}_0^*$	Fitted Frequency	Observed Frequency
ETA	0	35.49	35	AEV	8	0.61	2
TEA	2	12.87	11	TXV	8	0.61	2
ETV	3	7.75	13	VTE	8	0.61	2
EVA	3	7.75	9	ATV	8	0.61	1
ETX	3	7.75	9	TAV	8	0.61	1
EXA	3	7.75	7	TVX	8	0.61	1
TEV	5	2.81	7	VET	8	0.61	1
VEA	5	2.81	5	VEX	8	0.61	1
XEA	5	2.81	2	VTX	8	0.61	1
TVA	5	2.81	2	VXA	8	0.61	1
VTA	5	2.81	2	AET	8	0.61	0
TEX	5	2.81	1	AEX	8	0.61	0
XTA	5	2.81	1	ATE	8	0.61	0
TXA	5	2.81	0	ATX	8	0.61	0
EXV	6	1.69	3	TAE	8	0.61	0
EXT	6	1.69	2	TAX	8	0.61	0
EAT	6	1.69	1	TVE	8	0.61	0
EVT	6	1.69	1	TXE	8	0.61	0
EVX	6	1.69	0	XET	8	0.61	0
EAX	6	1.69	0	XEV	8	0.61	0
EAV	6	1.69	0	XTE	8	0.61	0
XTV	8	0.61	3	XVA	8	0.61	0

Table 1. Continued.

Partial Ranking	Distance to $\hat{\pi}_0^*$	Fitted Frequency	Observed Frequency	Partial Ranking	Distance to $\hat{\pi}_0^*$	Fitted Frequency	Observed Frequency
AXV	11	0.13	2	VAT	11	0.13	0
XAV	11	0.13	1	VAX	11	0.13	0
AVE	11	0.13	0	VXE	11	0.13	0
AVT	11	0.13	0	XAE	11	0.13	0
AVX	11	0.13	0	XAT	11	0.13	0
AXE	11	0.13	0	VXT	11	0.13	0
AXT	11	0.13	0	XVE	11	0.13	0
VAE	11	0.13	0	XVT	11	0.13	0

Explanation: Each student provided a partial ranking of the five words "emotion," "author," "experimentation," "thought," and "vigil," according to their degree of association with the word "obsession." The partial rankings above are encoded in the form: first choice, second choice, fifth choice. The following abbreviations are used: E = emotion, A = author, X = experimentation, T = thought, V = vigil.

(or $\hat{\pi}_0^*$ = ETA for short), corresponding to the fact that the model (and the data) are centered about this partial ranking. Meanwhile, the M.L.E. of the dispersion parameter λ is $\hat{\lambda} = 0.507$. This is somewhere in the middle of the range of possible values for $\hat{\lambda}$, yielding a model that is neither extremely peaked about $\hat{\pi}_0^*$, nor extremely dispersed.

The results in Table 1 were produced by a FORTRAN computer program, which first found the M.L.E.'s $\hat{\pi}_0^*$ and $\hat{\lambda}$, and then computed the fitted frequencies. Details are given in Appendix C.

A visual inspection of Table 1 indicates a reasonably good fit of the model to the data. We would like to confirm this visual impression by calculating a goodness-of-fit statistic, of the type discussed in Section 5:

$$X^2 = \sum_{i=1}^{M} \frac{(O_i - E_i)^2}{E_i} .$$

However, it seems inappropriate to use M = 60 cells for this calculation, that is, one cell for each possible partial ranking. The reason is that many of the possible partial rankings have very small fitted frequencies under the model, and this often gives rise to misleading values of X^2.

Therefore, some of the partial rankings were lumped together, into the same cell, before calculating X^2. In particular, this clustering was done according to distance from $\hat{\pi}_0^*$: all partial rankings at distance 5 from $\hat{\pi}_0^*$ were lumped together into a single cell, as were all partial rankings at distance 6 from $\hat{\pi}_0^*$, and all partial rankings at distance greater than 6. The results are shown in Table 2.

Again, a scan of the data indicates that the model fits well.

Table 2. Goodness-of-Fit of the Mallows Model to the Educational Testing Service Data.

Partial Ranking	Fitted Frequency E_i	Observed Frequency O_i
ETA	35.49	35
TEA	12.87	11
ETV	7.75	13
EVA	7.75	9
ETX	7.75	9
EXA	7.75	7
all rankings at distance 5 from ETA	22.50	20
all rankings at distance 6 from ETA	11.86	7
all rankings at distance >6 from ETA	16.28	19

$$X^2 = \frac{(O_i - E_i)^2}{E_i} = 7.03$$

This visual impression is reinforced by the fact that $X^2 = 7.03$. Indeed, if the model is valid, then X^2 is asymptotically distributed as a chi-squared statistic with 9-2=7 degrees of freedom, by the proposition of Section 5. The observed value $X^2 = 7.03$ is close to the median of this asymptotic distribution.

In summary, the extension of Mallows' model provides a nice parsimonious description of the E.T.S. data. The partial ranking $\hat{\pi}_0^*$ is, in some sense, the center of the distribution of responses. The observed frequencies of all other possible partial rankings tend to fall off exponentially, according to their distance from $\hat{\pi}_0^*$.

*   *   *

Finally, it is interesting to note that if we try to fit Mallows' model to the original full rankings, using Spearman's footrule metric F on S_5, a much worse fit is obtained. Indeed, if we cluster the full rankings according to the same rules as before, we obtain a value of X^2 which is the .996 quantile of its asymptotic distribution. Thus, in this example, the essential features of the data are captured better by the model which uses metrics on the coset space $S_5/S_1 \times S_1 \times S_2 \times S_1$.

B. Multidimensional Scaling for Partially Ranked Data

This section discusses how the metrics of Chapters III and IV can be utilized for the multidimensional scaling of partially ranked data. It also presents an example of this application of the metrics. This is a generalization of some work by Diaconis, who proposes a metric-based approach to the multidimensional scaling of fully ranked data [D2].

When one is confronted with a large data set, it is often almost impossible to get any preliminary impressions by just looking at the raw data. This can be especially true for partially ranked data, because each data point is a partial ranking and therefore represents an abundance of information. Thus, when a large number of people provide partial rankings of a large number of items, even relatively straightforward questions can be difficult to answer. For example: are the data points evenly spread out among all the possible partial rankings, or do they fall into one or more relatively disjoint clusters? Is there any noticeable difference between the partial rankings provided by two different groups of judges?

Multidimensional scaling is a statistical procedure which can be helpful for obtaining preliminary impressions of a large data set. In its most general form, multidimensional scaling operates on a collection of data points in an arbitrary metric space. It associates to each such data point a point in the plane, in such a way that nearby points in the metric space get mapped into nearby points in the plane. Thus, multidimensional scaling produces a collection of dots, and looking at these dots provides some visual information about the original data.

The "inner workings" of a multidimensional scaling computer program involve complex iterative optimization algorithms. These have been discussed by several authors [M2,G1,G3,K4,K5], and need not be presented here. What matters here is the fact that, to apply multidimensional scaling to partially ranked data, it must be shown that the partial rankings lie in a metric space. In other words, the _input_ to the multidimensional scaling program consists not only of the partial rankings themselves, but also of a set of _distances_ between the partial rankings.

Of course, these distances can now be obtained, using any of the metrics derived in Chapters III and IV.

1. An Example, Using Leann Lipps Birch's Cracker Preference Data

To illustrate multidimensional scaling for partially ranked data, using one of the metrics from Chapter III, consider the data in

Table 3, which was provided by Leann Lipps Birch [B3]. Twenty-two preschool boys, and sixteen mothers of preschool children, each tasted five types of crackers: cheese crackers, graham crackers, ritz crackers, saltines, and animal crackers. Each person then provided a partial ranking of the five crackers, by specifying his or her first, second, and third choices.

These partial rankings can be viewed as points in the coset space S_5/S_3. The distances between them can therefore be computed, using the Hausdorff extension R^* of Spearman's rho.

The entire collection of $\binom{38}{2}$ interpoint distances was fed into a standard multidimensional scaling program; the output is shown in Figure 3. Each triangle represents a partial ranking provided by one of the little boys, while each circle corresponds to one of the mothers' partial rankings. There are thus 22 triangles and 16 circles in all.

By looking at Figure 3, an interesting fact emerges. There seems to be a significant difference between the boys' partial rankings and the mothers'! In particular, it is possible to draw a straight line through the configuration of Figure 3, such that only two triangles lie on one side of the line, and only two circles lie on the other side.

One possible explanation for this phenomenon is suggested by examining more closely the raw data in Table 3. The mothers liked the saltines much more than the little boys. On the other hand, there was also one kind of cracker which the little boys liked much more than the mothers — namely, the animal crackers.

Table 3. Partial Rankings of Five Types of Crackers.

BOYS' RANKINGS	MOTHERS' RANKINGS
ACS	CRA
GCA	SRG
ACG	CSA
CAG	CSA
CGA	SRA
ARC	SCR
CSA	SCG
SCR	GAR
AGC	SAR
ARG	CSA
AGC	RSC
ACS	RAG
GRA	SCG
CGA	SAR
ACS	GAS
CGS	SCA
ARC	
ACG	
RAC	
AGC	
ACG	
CAG	

Explanation: Partial rankings are encoded in the form: first choice, second choice, third choice. The following abbreviations are used: A = animal crackers, C = cheese crackers, G = graham crackers, R = ritz crackers, S = saltines.

Figure 3. Multidimensional Scaling for the Cracker Preference Data.

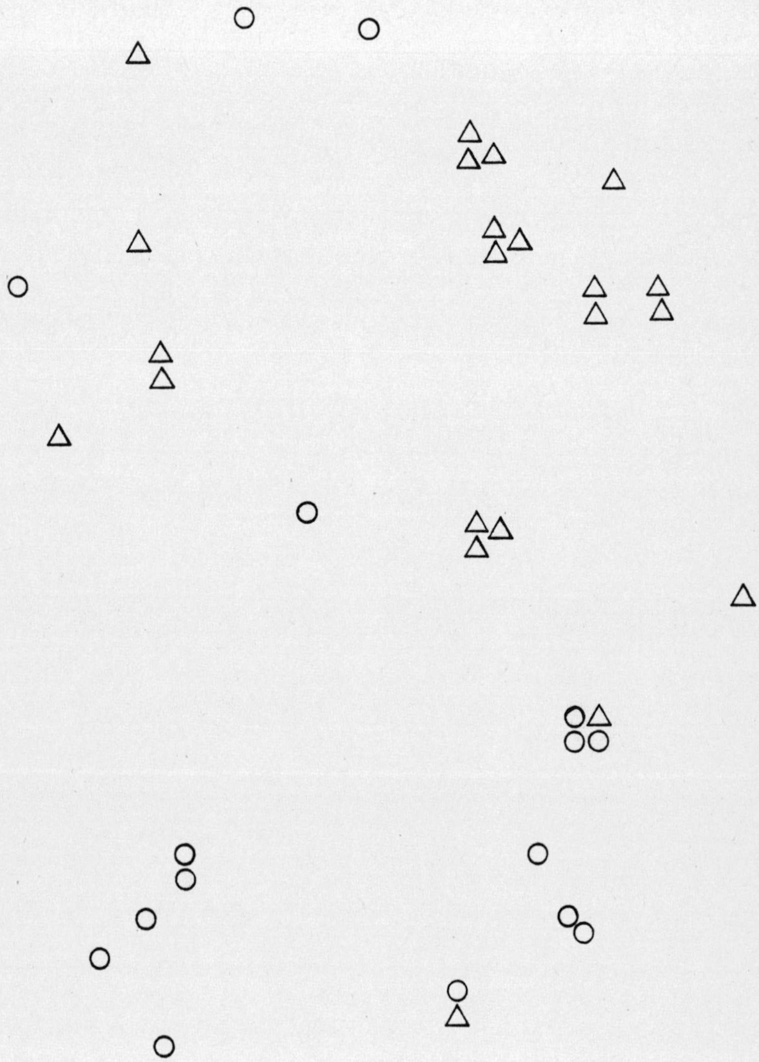

Explanation: The interpoint distances between the partial rankings in Table 3 were calculated, using the Hausdorff metric R^*, and were then fed into a standard multidimensional scaling program. The output is displayed above: each triangle represents one of the little boys' rankings, while each circle represents one of the mothers' rankings.

In sum, when used in conjunction with the metric R^*, multidimensional scaling provides a nice picture of this partially ranked data set, and suggests a real difference between the boys' cracker preferences and the mothers'.

C. Two-Sample Problems for Partially Ranked Data

The example of the last section points to the following general statistical problem: suppose there are two distinct subpopulations of judges, and each judge gives a partial ranking of the same set of items. Is there a "statistically significant" difference between the partial rankings produced by the two subpopulations of judges?

In the example of the previous section, the exploratory multidimensional scaling technique <u>suggested</u> that there was a significant difference, but it did not <u>reinforce</u> this impression by calculating a "p-value" or "significance level" for the observed data. The purpose of the present section is to approach this two-sample problem from a formal, hypothesis-testing point of view, and to illustrate how metrics can be useful for resolving the problem.

Two separate ways of attacking the problem are discussed: the first is based on the minimal spanning tree of the data, and the second is based on the nearest-neighbors graph. Both methodologies were originally introduced by Jerome Friedman and Lawrence Rafsky for data taking values in an arbitrary metric space [F3,F4], and were specialized to the case of fully ranked data by Diaconis [D2]. To apply these methodologies to partially ranked data, we must use the metrics on partial rankings that were derived in Chapters III and IV.

The two approaches are illustrated on Leann Lipps Birch's cracker preference data, described in the previous section.

1. A Two-Sample Test Based on the Minimal Spanning Tree

To fix notation, suppose that there is an overall population of J judges, and that J_1 of these judges belong to the first subpopulation, while the remaining $J_2 = J - J_1$ judges comprise the second subpopulation. The null hypothesis H_0 asserts that there is no significant difference between the partial rankings provided by the two subpopulations, while the alternative hypothesis H_1 states that there _is_ a significant difference.

To test these hypotheses, fix a metric on partial rankings. Regard the J partial rankings as J points in a metric space, and calculate all the $\binom{J}{2}$ distances between pairs of partial rankings. Use these distances to construct the minimal spanning tree of the data points — that is, the shortest graph which has a path between any two distinct points. Let X be the number of edges in the minimal spanning tree which join data points (i.e., partial rankings) from _different_ subpopulations of judges.

Suppose now that there is a significant difference between the two sets of partial rankings. Then the two sets of points will tend to fall into two separate clusters, so X, the number of "cross-overs," should be relatively small. Therefore, we want to reject H_0 for small values of X, and to accept H_0 for large values. Friedman and Rafsky determine the asymptotic distribution of X under H_0, so that we can find an approximate p-value for any observed value of X.

* * *

The foregoing methodology will now be applied to Leann Lipps Birch's cracker preference data. To test for significant differences between the boys' partial rankings and the mothers', the interpoint distances — that is, the distances between pairs of partial rankings — were calculated, using the Hausdorff metric R^* on S_5/S_3. A minimal spanning tree of the data points was then formed.

It is impossible to draw a completely accurate picture of this data set, since the data points belong to the abstract metric space S_5/S_3, rather than to the euclidean plane $\mathbb{R}^2$. Nonetheless, a pictorial representation of the minimal spanning tree is attempted in Figure 4. In Figure 4, each lower case letter corresponds to one of the little boys' partial rankings, while each capital letter represents one of the mothers' partial rankings. For both lower case and capital letters, the usual alphabetical ordering of these letters corresponds to the order in which the rankings were presented in Table 3 of Section B. The lines in Figure 4 represent the edges of the minimal spanning tree, while the numbers are the <u>squares</u> of the interpoint distances, calculated according to the metric R^*.

The minimal spanning tree in Figure 4 has 11 edges which join capital letters to lower case letters, so $X = 11$. However, in this example, we cannot immediately apply Friedman and Rafsky's arguments to obtain the corresponding p-value. The reason is that the minimal spanning tree for this data set is not unique. There are other minimal spanning trees which lead to other values of X. Since the

Figure 4. A Minimal Spanning Tree for the Cracker Preference Data.

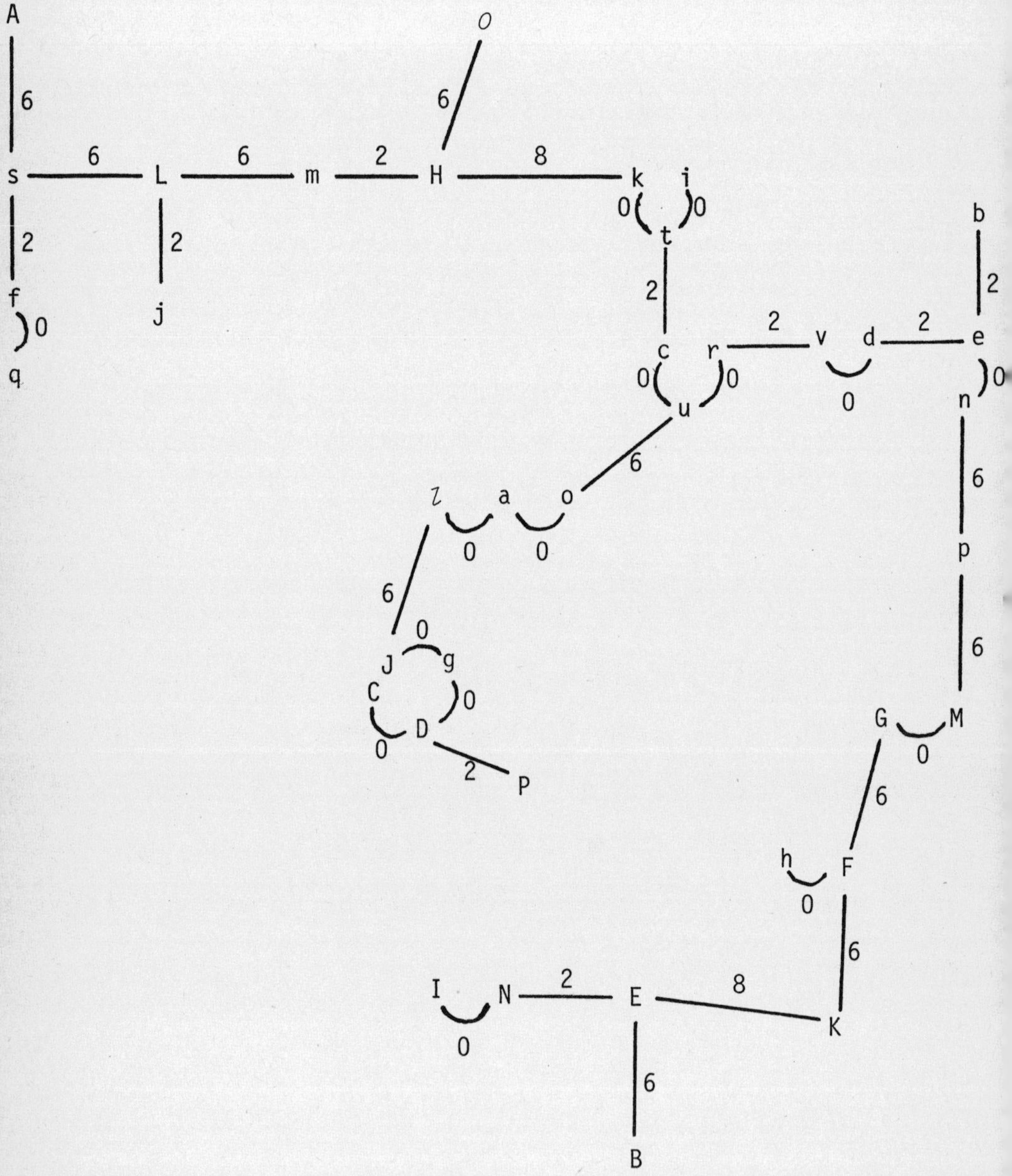

Explanation: Each lower case letter represents one of the little boys' partial rankings, while each capital letter represents one of the mothers' rankings. The lines are the edges of the minimal spanning tree, and the numbers are the squares of the interpoint distances, calculated according to the Hausdorff metric R^*.

minimal spanning tree is not uniquely defined, neither is the statistical test of H_0 versus H_1.

One way to get around this difficulty is to consider a slightly different graph from the one in Figure 4 — namely, the union of all minimal spanning trees for this data set. This graph has the advantage of being uniquely determined by the data. It turns out to have a total of t = 102 edges, of which X = 31 are "cross-overs."

Is this "statistically significant" evidence of a difference between the two groups of judges? To calculate the corresponding p-value, we apply the results in Friedman and Rafsky's article, to obtain the asymptotic distribution of X under H_0. These arguments are valid for an arbitrary graph having J nodes, J_1 of which are from the first subpopulation, $J_2 = J - J_1$ of which are from the second subpopulation, and having t edges, X of which are "cross-overs" between the two types of nodes. Under the null hypothesis H_0, each edge has a probability

$$p = \frac{2\ J_1 J_2}{J(J-1)} = \frac{2(22)(16)}{(38)(37)} = .5007$$

of joining two nodes from the different subpopulations. Therefore, the expected value of X is

$$E(X|t) = pt = (.5007)(102) = 51.07.$$

To find the variance of X under H_0, Friedman and Rafsky argue that one must condition not only on t, but also on c, the number of edge pairs in the graph which share a common node. They then show that

$$\mathrm{Var}(X|t,c) = p\,\{t+c+\frac{2(J_1-1)(J_2-1)}{(J-2)\,(J-3)}\;[t(t-1) - 2c] - pt^2\}.$$

In the case at hand, c = 514, so $\mathrm{Var}(X|t,c)$ works out to 43.6, and the conditional standard deviation is thus

$$\sigma(X|t,c) = (43.6)^{1/2} = 6.6.$$

Finally, assuming c is O(J), Friedman and Rafsky prove that the (conditional) distribution of X is asymptotically normal under H_0.

Putting together all of these pieces, we can now find an approximate p-value for the observed data X = 31, t = 102, c = 514. Let Φ be the standard normal cumulative distribution function. The p-value is

$$\Phi(\frac{X-E(X|t,c)}{\sigma(X|t,c)}) = \Phi(\frac{X-E(X|t)}{\sigma(X|t,c)}) = \Phi(-3.04) = 0.0012.$$

By any reasonable standard, this is "statistically significant" evidence for concluding that there is a real difference between the two groups of judges.

2. A Two-Sample Test Based on the Nearest Neighbors Graph

The union of all minimal spanning trees, considered in the previous section, is somewhat cumbersome for this data set. There are several partial rankings which occur more than once in the data, and these lead to quite large values of t and c.

Since Friedman and Rafsky's arguments are valid for arbitrary graphs, it may therefore be more convenient to consider the "nearest neighbors graph" for this data set. It is formed by linking each data point to its "nearest neighbor," as determined by the interpoint distances under the metric R^*. If there is more than one nearest neighbor for any data point, then all of the corresponding edges are included in the nearest neighbors graph.

The nearest neighbors graph for the cracker preference data is shown in Figure 5. It has a total of t = 42 edges, of which X = 11 link rankings from the two different groups of judges. Also, there are c = 68 edge pairs which share a common node.

Under the null hypothesis H_0, each edge has a probability

$$p = \frac{2\ J_1J_2}{J(J-1)} = .5007$$

of joining two nodes from the different groups. Friedman and Rafsky's equations yield

$$E(X|t) = pt = (.5007)(42) = 21.03$$

and

$$Var(X|t,c) = p\{t+c+\frac{2(J_1-1)(J_2-1)}{(J-2)(J-3)}[t(t-1)-2c] - pt^2\} = 9.89,$$

and thus

$$\sigma(X|t,c) = 3.14.$$

By the asymptotic normality result cited in Section 1, the p-value corresponding to the observed data is

$$\Phi(\frac{X-E(X|t,c)}{\sigma(X|t,c)}) = \Phi(\frac{X-E(X|t)}{\sigma(X|t,c)}) = \Phi(-3.19) = 0.00071.$$

Figure 5. The Nearest Neighbors Graph for the Cracker Preference Data.

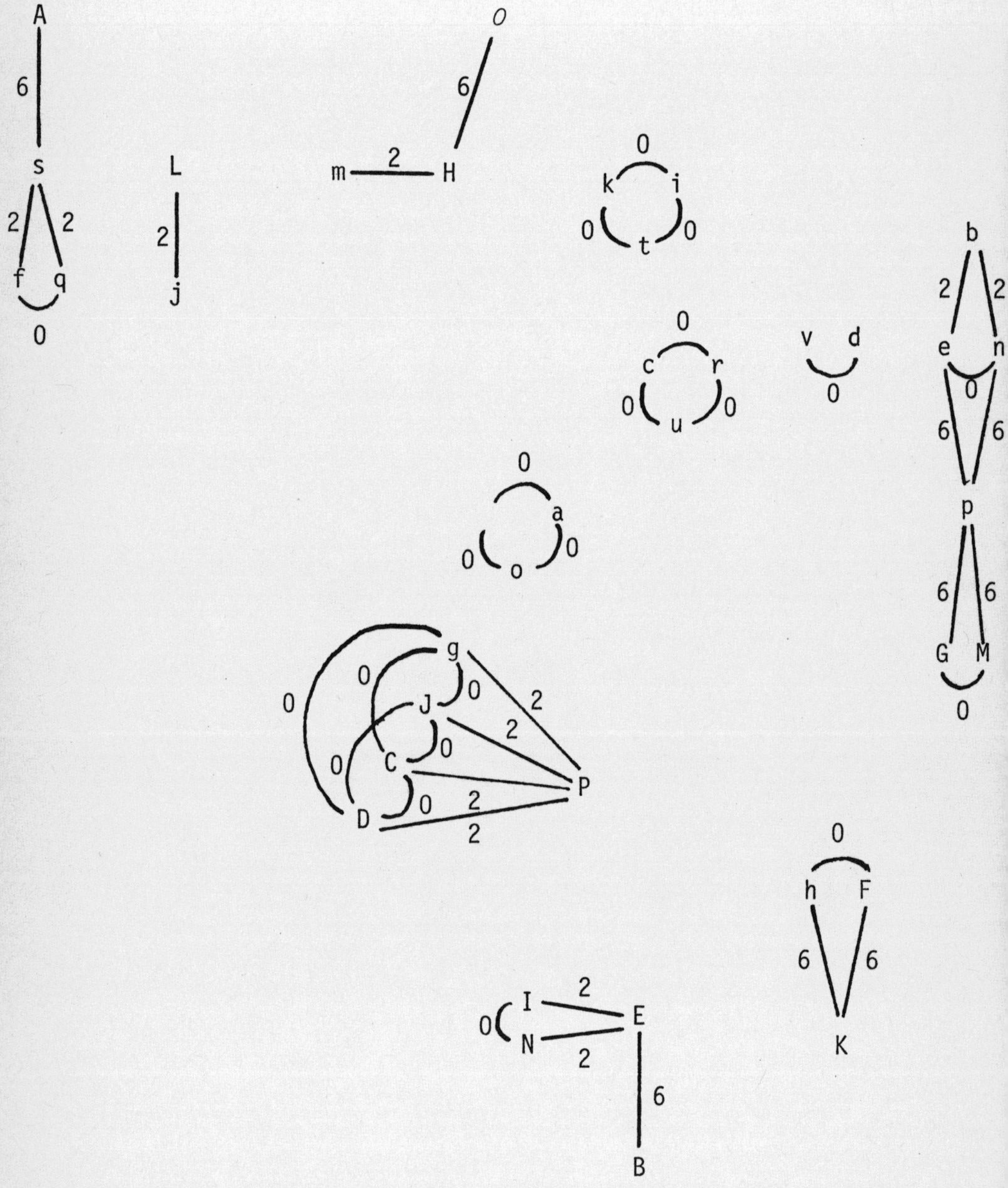

<u>Explanation</u>: As in Figure 4, each lower case letter represents one of the little boys' rankings, and each capital letter represents one of the mothers' rankings. The lines are the edges of the nearest neighbors graph, and the numbers are the squares of the interpoint distances, calculated according to the Hausdorff metric R^*.

As before, we conclude that there is, indeed, a significant difference between the little boys' rankings and the mothers'.

APPENDIX A - THE EXISTENCE OF FIXED VECTORS

If the group G, the subgroup K, the representation ρ, and the fixed vector $\underset{\sim}{v}$ satisfy certain conditions, then we saw in Section III.C that the fixed vector method produces a right-invariant metric on the coset space G/K. Moreover, we found that the fixed vector metric has a surprisingly nice statistical interpretation, in a few special cases.

For the fixed vector construction to be useful, there must <u>exist</u> a vector $\underset{\sim}{v}$ fixed by K under ρ^T. That is, there must exist $\underset{\sim}{v}\in\mathbb{R}^n$ such that

$$\rho(k)^T\underset{\sim}{v} = \underset{\sim}{v} \qquad \forall k\in K.$$

In this appendix, we address the question of when such fixed vectors exist, and indeed we develop a general formula for the <u>dimension</u> of the space of fixed vectors.

In so doing, we will make contact with the elaborate machinery for studying group representations, developed by many mathematicians. It is hoped that this additional material will make the fixed vector metric more meaningful, by exhibiting it as part of a larger framework. We outline here the relevant theory, often without proofs; for further details the reader may consult Serre [S2].

Let Δ be the dimension of the space of vectors fixed by K under ρ^T:

$$\Delta = \dim\{\underset{\sim}{v}\in\mathbb{R}^n : \rho(k)^T\underset{\sim}{v} = \underset{\sim}{v} \quad \forall k\in K\}.$$

To evaluate Δ, we proceed in a step-by-step fashion, and state our "grand conclusion" at the end.

Step 1: We assume that ρ is a unitary representation, i.e., that $\rho(g)$ is a unitary matrix for all $g \in G$. This is not a very restrictive assumption, since any finite-dimensional representation is equivalent to a unitary representation [S2, p. 6]. As a consequence, note that

$$\underset{\sim}{v} \text{ is fixed by } K \text{ under } \rho^T \Longleftrightarrow \underset{\sim}{v} \text{ is fixed by } K \text{ under } \rho,$$

because

$$\rho(k)^T \underset{\sim}{v} = \rho(k)^{-1} \underset{\sim}{v} = \rho(k^{-1}) \underset{\sim}{v} \qquad \forall \underset{\sim}{v} \in \mathbb{R}^n \quad \forall k \in K.$$

Hence, Δ can be re-expressed as the dimension of $\rho(K)$-fixed vectors:

$$\Delta = \dim \{ \underset{\sim}{v} \in \mathbb{R}^n : \rho(k) \underset{\sim}{v} = \underset{\sim}{v} \qquad \forall k \in K \}.$$

Step 2: When ρ is restricted to the subgroup K, it obviously induces a representation of K, denoted $\rho{\downarrow}K$. Moreover, the dimension of $\rho(K)$-fixed vectors is the dimension of the largest subspace of $\mathbb{R}^n$, such that $\rho{\downarrow}K$ acts as the identity on this subspace.

Group theorists have another way of expressing the same idea, which makes use of the fact that the representation $\rho{\downarrow}K$ can be decomposed as a direct sum of irreducible representations. Indeed, we can write $\mathbb{R}^n$ as a direct sum

$$V_1 \oplus \ldots \oplus V_m$$

of non-trivial subspaces, such that each V_i is invariant under the action of $\rho{\downarrow}K$, and such that no non-trivial subspace of V_i is invariant. That is,

$$\rho(k)V_i \subset V_i \quad \forall k\varepsilon K \quad \forall i=1,\ldots,m,$$

and

$$\rho(K)W \not\subset W \text{ for all subspaces } W \text{ such that } 0 \neq W \subsetneq V_i.$$

We can then write $\rho{\downarrow}K$ as the direct sum of its restrictions to these subspaces:

$$\rho{\downarrow}K = (\rho{\downarrow}K)|_{V_1} \oplus \ldots \oplus (\rho{\downarrow}K)|_{V_m}.$$

The preceding direct sum decomposition yields the following alternative description of the dimension of $\rho(K)$-fixed vectors. Let ϕ be the <u>trivial representation</u> of the subgroup K. Thus, for each $k\varepsilon K$, $\phi(k)$ is the identity mapping: $\mathbb{R} \to \mathbb{R}$. The <u>multiplicity</u> of ϕ in $\rho{\downarrow}K$ is the number of times that ϕ appears in the direct sum decomposition of $\rho{\downarrow}K$:

$$\text{multiplicity of } \phi \text{ in } \rho{\downarrow}K = \#\{i=1,\ldots,m\colon (\rho{\downarrow}K)|_{V_i} \cong \phi\}.$$

It is the dimension of the largest subspace of $\mathbb{R}^n$, such that $\rho{\downarrow}K$ is the identity on this subspace. Therefore, the multiplicity of ϕ in $\rho{\downarrow}K$ equals the dimension of $\rho(K)$-fixed vectors:

$$\Delta = \text{multiplicity of } \phi \text{ in } \rho{\downarrow}K.$$

Step 3: The representations ϕ and $\rho{\downarrow}K$, of the subgroup K, can be "lifted" to representations of the full group G. These "liftings" are called induced representations, and are described more completely in Serre [S2, Section 3.3]. The representation induced by $\rho{\downarrow}K$ is just the original representation ρ. Meanwhile, by Exercise 3.5 of Serre [S2, p. 31], the representation η induced by ϕ is as follows.

Let L(G/K) denote the vector space of all real-valued functions on the coset space G/K. Then, for each $g \in G$, $\eta(g)$ is the linear transformation

$$\eta(g): L(G/K) \to L(G/K)$$

given by

$$\eta(g)f(x) = f(xg) \qquad \forall x \in G/K \qquad \forall f \in L(G/K).$$

Step 4: The Frobenius Reciprocity Theorem [S2, Section 7.2] states that "multiplicities are preserved by lifting to induced representations." In other words, the multiplicity of ϕ in $\rho{\downarrow}K$ equals the multiplicity of η in ρ. We therefore conclude:

$$\Delta = \text{multiplicity of } \eta \text{ in } \rho.$$

Step 5: Finally, group theorists have a handy way of computing multiplicities, such as the one we derived in Step 4. The character χ_ρ of the representation ρ is the function

$$\chi_\rho: \ G \to \mathbb{R}$$

given by

$$\chi_\rho(g) = \mathrm{Tr}(\rho(g)) \qquad \forall g \varepsilon G.$$

Here $\mathrm{Tr}(\rho(g)) = \sum_{i=1}^{n} \rho(g)_{ii}$ denotes the trace of the matrix $\rho(g)$; it is independent of the choice of basis.

Similarly, the character χ_η of η is given by

$$\chi_\eta(g) = \mathrm{Tr}(\eta(g)) \qquad \forall g \varepsilon G.$$

The <u>inner product</u> of the characters χ_ρ and χ_η is

$$(\chi_\rho|\chi_\eta) = \frac{1}{\#G} \sum_{g\varepsilon G} \mathrm{Tr}(\rho(g))\mathrm{Tr}(\eta(g)).$$

A basic theorem [S2, p. 16] asserts that the multiplicity of η in ρ is the same as the inner product $(\chi_\rho|\chi_\eta)$.

Putting all of Steps 1 through 5 together, we arrive at the following

<u>Conclusion</u>: The dimension Δ of the space of vectors fixed by K under ρ^T equals the inner product $(\chi_\rho|\chi_\eta)$:

$$\Delta = (\chi_\rho|\chi_\eta) = \frac{1}{\#G} \sum_{g\varepsilon G} \mathrm{Tr}(\rho(g))\mathrm{Tr}(\eta(g)),$$

where η is the induced representation of G described in Step 3. In particular, fixed vectors exist if and only if $(\chi_\rho|\chi_\eta) > 0$.

* * *

<u>An Example</u>: To illustrate the foregoing theory, we return to Example 2 of Section III.C, where

$$G = S_n, \; K = S_{n-k} = \{\pi\varepsilon S_n : \pi(i) = i \quad \forall i = 1,\ldots,k\},$$

and where ρ = the n-dimensional permutation representation. In this case, it is easy to compute the dimension of the space of fixed vectors, by elementary arguments. Indeed, as shown in Section III.C, $\rho(\pi)^T$ simply permutes the components of any vector $\underset{\sim}{w}$:

$$(\rho(\pi)^T \underset{\sim}{w})_i = w_{\pi(i)} \qquad \forall\pi\varepsilon S_n \qquad \forall \underset{\sim}{w}\varepsilon \mathbb{R}^n.$$

Therefore,

$$\begin{array}{l}\underset{\sim}{v} = (v_1,\ldots,v_n)^T \text{ is} \\ \text{fixed by } S_{n-k} \text{ under } \rho^T\end{array} \Longleftrightarrow v_{k+1} = v_{k+2} = \cdots = v_n,$$

and the dimension of fixed vectors is clearly k+1.

We now derive the same result, using the general theory developed in this appendix. The dimension Δ of fixed vectors is

$$\Delta = (\chi_\rho | \chi_\eta) = \frac{1}{n!} \sum_{\pi\varepsilon S_n} \mathrm{Tr}(\rho(\pi))\mathrm{Tr}(\eta(\pi)),$$

where

$$\mathrm{Tr}(\rho(\pi)) = \#\text{ of fixed points of } \pi,$$

and where

$$\begin{aligned}\mathrm{Tr}(\eta(\pi)) &= \#\ \{x\varepsilon S_n/S_{n-k} : x\pi = x\} \\ &= \frac{1}{(n-k)!}\ \#\ \{\sigma\varepsilon S_n : \sigma\pi\varepsilon S_{n-k}\sigma\} \\ &= \frac{1}{(n-k)!}\ \#\ \{\sigma\varepsilon S_n : \sigma\pi\sigma^{-1}\varepsilon S_{n-k}\}.\end{aligned}$$

Now for any $\sigma \varepsilon S_n$,

$$\sigma\pi\sigma^{-1} \varepsilon S_{n-k} \Longleftrightarrow \{1,\ldots,k\} \subset \{\text{fixed points of } \sigma\pi\sigma^{-1}\}$$

$$\Longleftrightarrow \{\sigma^{-1}(1),\ldots,\sigma^{-1}(k)\} \subset \{\text{fixed points of } \pi\}.$$

If π has exactly i fixed points, then the number of permutations σ satisfying the preceding condition is clearly 0 if $i < k$, while if $i \geq k$ this number is $\binom{i}{k}k!(n-k)!$ This is because there are $\binom{i}{k}$ ways of choosing the unordered set $\{\sigma^{-1}(1),\ldots,\sigma^{-1}(k)\}$. For each of these choices, there are k! possibilities for the <u>ordered</u> k-tuple $(\sigma^{-1}(1),\ldots,\sigma^{-1}(k))$, and for each of these, there are (n-k)! ways of choosing the ordered (n-k)-tuple $(\sigma^{-1}(k+1),\ldots,\sigma^{-1}(n))$.

Putting all of this together, we obtain

$$\Delta = \frac{1}{n!(n-k)!} \sum_{i=k}^{n} \#\{\pi\varepsilon S_n\colon \pi \text{ has exactly } i \text{ fixed points}\}\cdot i\cdot\binom{i}{k}k!(n-k)!$$

Furthermore, for each $i = k,\ldots,n$,

$$\#\{\pi\varepsilon S_n\colon \pi \text{ has exactly } i \text{ fixed points}\}$$

$$= \binom{n}{i}\ \#\{\pi\varepsilon S_{n-i}\colon \pi \text{ has no fixed points}\}$$

$$= \binom{n}{i}(n-i)! \sum_{j=0}^{n-i} \frac{(-1)^j}{j!}$$

by elementary combinatorial arguments. Hence

$$\Delta = \frac{1}{n!(n-k)!} \sum_{i=k}^{n} \binom{n}{i}(n-i)!\left[\sum_{j=0}^{n-i} \frac{(-1)^j}{j!}\right] i\binom{i}{k}k!(n-k)!$$

After cancelling common factors, this yields

$$\Delta = \sum_{i=k}^{n} \frac{i}{(i-k)!} \sum_{j=0}^{n-i} \frac{(-1)^j}{j!}$$

$$= \sum_{i=0}^{n-k} \frac{i+k}{i!} \sum_{j=0}^{n-k-i} \frac{(-1)^j}{j!}$$

$$= \sum_{i=1}^{n-k} \sum_{j=0}^{n-k-i} \frac{(-1)^j}{(i-1)!j!} + k \sum_{i=0}^{n-k} \sum_{j=0}^{n-k-i} \frac{(-1)^j}{i!j!}$$

$$= \sum_{i=0}^{n-k-1} \sum_{j=0}^{n-k-1-i} \frac{(-1)^j}{i!j!} + k \sum_{i=0}^{n-k} \sum_{j=0}^{n-k-i} \frac{(-1)^j}{i!j!} .$$

In the first double sum above, we are summing over all non-negative pairs (i,j) such that $0 \leq i+j \leq n-k-1$, while in the second double sum, we are summing over all non-negative pairs (i,j) such that $0 \leq i+j \leq n-k$. Letting $s = i+j$ in each instance, we can re-express these double sums, yielding

$$\Delta = \sum_{s=0}^{n-k-1} \sum_{i=0}^{s} \frac{(-1)^{s-i}}{i!(s-i)!} + k \sum_{s=0}^{n-k} \sum_{i=0}^{s} \frac{(-1)^{s-i}}{i!(s-i)!}$$

$$= \sum_{s=0}^{n-k-1} \frac{1}{s!} \sum_{i=0}^{s} \binom{s}{i}(-1)^{s-i} + k \sum_{s=0}^{n-k} \frac{1}{s!} \sum_{i=0}^{s} \binom{s}{i}(-1)^{s-i}$$

$$= \sum_{s=0}^{n-k-1} \frac{1}{s!} (1-1)^s + k \sum_{s=0}^{n-k} \frac{1}{s!} (1-1)^s$$

$$= 1+k\cdot 1$$

$$= k+1,$$

which agrees with our previous result. Thus, the intricate representation theory "works" in this example, because it produces the right answer for the dimension of fixed vectors.

APPENDIX B - FORTRAN SUBROUTINES FOR EVALUATING THE METRICS ON S_n/S_{n-k} AND S_n/S

1. The Subroutine METRIC

This subroutine (pages 143-148), written in standard FORTRAN, computes the distance between two partial rankings with respect to a specified metric on S_n/S_{n-k}. Thus, the partial rankings should be of the type considered in Chapter III (where each judge lists his k favorite items out of n), and the distance between them is calculated according to any of the Hausdorff or fixed vector metrics of Sections III.B and III.C.

Input Parameters: The input parameters include k and n, discussed above, as well as two k-vectors i and j, which represent the two partial rankings. The vector i lists the first judge's k favorite items in order, while the vector j does the same for the second judge. More formally, if the two judges' partial rankings are the right cosets $S_{n-k}\pi$ and $S_{n-k}\sigma$, respectively, where $\pi,\sigma\varepsilon S_n$, then

$$i = (\pi^{-1}(1),\ldots,\pi^{-1}(k))$$

and

$$j = (\sigma^{-1}(1),\ldots, \sigma^{-1}(k)).$$

The input parameter metnum specifies which metric is to be used. The desired distance is then computed by calling an appropriate subroutine of METRIC, in accordance with Table 4.

Table 4. Subroutines of METRIC, the Associated Metrics on S_n/S_{n-k}, and the Corresponding Values of the Input Parameter metnum.

Subroutine	Metric	Value of metnum
TAUSTAR	T^*	1
RHOSTAR	R^*	2
RHOFV	R_{fv}	3
FOOTSTAR	F^*	4
FOOTFV	F_{fv}	5
CAYSTAR	C^*	6
HAMSTAR	H^*	7
ULAMSTAR	U^*	8

The final inputs are the two n-vectors iw and jw. The purpose of these vectors is to provide additional storage space necessary for computing C^* and U^*. Therefore, the input values of these vectors need not be specified.

Output Parameters: The output parameter m is the value of the metric. The only exceptions to this are for the metrics R_{fv} and R^*, for which the output parameter m equals $2R_{fv}^2$ and R^{*2}, respectively. The reason for these transformations is that the metrics R_{fv} and R^* are not always integer valued, whereas the output parameter m must be.

The output parameter max gives the maximum possible value of the metric:

$$\max = \max\{d^*(\pi^*,\sigma^*): \pi^*,\sigma^* \varepsilon S_n/S_{n-k}\}.$$

(Again, R_{fv} and R^* are replaced here by $2R_{fv}^2$ and R^{*2}, so that max is always integer valued.)

Algorithms Used: From the formulas of Sections III.B and III.C, it was straightforward to implement the various metrics in their respective subroutines TAUSTAR, RHOSTAR, RHOFV, etc. — with the sole exception of the metric U^*, discussed below.

By the theorem of Section III.B, to evaluate U^* one must calculate the ordinary Ulam's distances $U(\alpha,id)$ and $U(\beta,id)$, where α and β are certain special full permutations. It is an easy programming task to construct α and β; the challenge arises in trying to compute $U(\alpha,id)$ and $U(\beta,id)$ efficiently.

Thus, the most interesting part of the subroutine ULAMSTAR is the subroutine ULAM, which evaluates the ordinary Ulam's distance $U(\tau,id)$ for an arbitrary full permutation $\tau\varepsilon S_n$. The underlying algorithm for computing $U(\tau,id)$ is due to Peter H. Sellers [S1], and is very efficient, requiring only order n^2 steps:

Define the array (d_{ij}) for $i,j = 0,1,\ldots,n$ by setting:

$$d_{i0} = i \quad \text{for all} \quad i = 0,1,\ldots,n$$

$$d_{0j} = j \quad \text{for all} \quad j = 0,1,\ldots,n$$

and by defining the remaining d_{ij} inductively, as follows:

$$d_{ij} = \begin{cases} \min\{d_{i-1,j} + 1,\ d_{i,j-1} + 1\} & \text{if } \tau^{-1}(i) = j \\ \min\{d_{i-1,j-1},\ d_{i-1,j} + 1,\ d_{i,j-1} + 1\} & \text{if } \tau^{-1}(i) \neq j. \end{cases}$$

Then $U(\tau,id) = d_{nn}/2$, by a special case of Sellers' Theorem 2 [S1, p. 790].

The subroutine ULAM is a straightforward implementation of Sellers' algorithm for computing $U(\tau,id)$. The input parameters include n, as well as the n-vector $k = (\tau^{-1}(1),\dots, \tau^{-1}(n))$. The input vector l provides necessary storage space: after the i-th run through the "do 30" loop, l stores the values $d_{i1},\dots,d_{in}$. Finally, the output parameter m is the value of Ulam's metric:

$$m = d_{nn}/2 = U(\tau,id).$$

```
C     THE SUBROUTINE METRIC
C
C
C     THIS SUBROUTINE CALCULATES THE DISTANCE BETWEEN TWO
C      PARTIAL RANKINGS WITH RESPECT TO A SPECIFIED METRIC
C      ON Sn/Sn-k.
C
      subroutine METRIC(metnum, i, j, iw, jw, k, n, m, max)
      integer i(k), j(k), iw(n), jw(n)
      l=metnum
      if (l .eq. 1) call TAUSTAR(i, j, k, n, m, max)
      if (l .eq. 2) call RHOSTAR(i, j, k, n, m, max)
      if (l .eq. 3) call RHOFV(i, j, k, n, m, max)
      if (l .eq. 4) call FOOTSTAR(i, j, k, n, m, max)
      if (l .eq. 5) call FOOTFV(i, j, k, n, m, max)
      if (l .eq. 6) call CAYSTAR(i, j, iw, k, n, m, max)
      if (l .eq. 7) call HAMSTAR(i, j, k, n, m, max)
      if (l .eq. 8) call ULAMSTAR(i, j, iw, jw, k, n, m, max)
      return
      end
C
C
      subroutine TAUSTAR(i, j, k, n, m, max)
      integer i(k), j(k)
      n1=0
      n2=0
      n3=k*(k+1)
      do 10 m1=1, k
      do 10 m2=1, k
      if (i(m1) .ne. j(m2)) go to 10
      n1=n1+1
      if (m1 .eq. 1) go to 10
      if (m2 .eq. k) go to 10
      do 10 m3=1, m1-1
      do 10 m4=m2+1, k
      if (i(m3) .eq. j(m4)) n2=n2+1
   10 continue
      do 20 m1=1, k
      do 20 m2=1, k
      if (i(m1) .eq. j(m2)) n3=n3-m1-m2
   20 continue
      l=k-n1
      m=n2+l*(n+k)-(l*(l-1))/2-n3
      max=k*n-(k*(k+1))/2
      return
      end
C
C
      subroutine RHOSTAR(i, j, k, n, m, max)
      integer i(k), j(k)
      n1=0
      n2=0
      n4=0
```

```
      n5=0
      ni=0
      nii=0
      niii=0
      niv=0
      do 30 m1=1,k
      n3=0
      do 20 m2=1,k
      if (i(m1) .ne. j(m2)) go to 10
      n1=n1+1
      n2=n2+(m1-m2)**2
      go to 30
   10 n3=n3+1
      if (n3 .lt. k) go to 20
      n4=n4+1
      ni=ni+(n+1-n4-m1)**2
      niii=niii+(k+n4-m1)**2
   20 continue
   30 continue
      do 50 m2=1,k
      n3=0
      do 40 m1=1,k
      if (i(m1) .eq. j(m2)) go to 50
      n3=n3+1
      if (n3 .lt. k) go to 40
      n5=n5+1
      nii=nii+(k+n5-m2)**2
      niv=niv+(n+1-n5-m2)**2
   40 continue
   50 continue
      l=k-n1
      na=ni+nii
      if (niii+niv .gt. na) na=niii+niv
      m=n2+na+(n-k-1)*(l**2)
      max=k*n*(n-k)+(k*(k+1)*(k-1))/3
      return
      end
C
C
      subroutine RHOFV(i,j,k,n,m,max)
      integer i(k),j(k)
      n1=0
      n2=0
      do 10 m1=1,k
   10 n2=n2+(n+k+1-2*m1)**2
      do 20 m1=1,k
      do 20 m2=1,k
      if (i(m1) .ne. j(m2)) go to 20
      n1=n1+2*((m1-m2)**2)
      n3=2.*(.5*float(n+k+1)-float(m1))**2.
     X   +2.*(.5*float(n+k+1)-float(m2))**2.+.1
      n2=n2-n3
   20 continue
```

```
      m=n1+n2
      max=0
      if (2*k .gt. n) go to 40
      do 30 l=1,k
   30 max=max+(n+k+1-2*l)**2
      go to 70
   40 do 50 l=k+1,n
   50 max=max+(2*l+k-n-1)**2
      do 60 l=n-k+1,k
   60 max=max+2*(n+1-2*l)**2
   70 continue
      return
      end
c
c
      subroutine FOOTSTAR(i,j,k,n,m,max)
      integer i(k),j(k)
      n1=0
      n2=0
      n3=k*(k+1)
      do 10 m1=1,k
      do 10 m2=1,k
      if (i(m1) .ne. j(m2)) go to 10
      n1=n1+1
      nabs=m1-m2
      if (m2 .gt. m1) nabs=-nabs
      n2=n2+nabs
      n3=n3-m1-m2
   10 continue
      l=k-n1
      m=l*(2*n+1-l)+n2-n3
      max=2*k*(n-k)
      return
      end
c
c
      subroutine FOOTFV(i,j,k,n,m,max)
      integer i(k),j(k)
      n1=0
      n2=0
      n3=k*(k+1)
      do 10 m1=1,k
      do 10 m2=1,k
      if (i(m1) .ne. j(m2)) go to 10
      n1=n1+1
      nabs=m1-m2
      if (m2 .gt. m1) nabs=-nabs
      n2=n2+nabs
      n3=n3-m1-m2
   10 continue
      l=k-n1
      m=l*(n+k+1)+n2-n3
      max=0
```

```
      if (2*k .gt. n) go to 30
      do 20 l=1,k
   20 max=max+(n+k+1-2*l)
      go to 60
   30 do 40 l=k+1,n
   40 max=max+2*l+k-n-1
      do 50 l=n-k+1,k
      nabs=n+1-2*l
      if (2*l .gt. n+1) nabs=-nabs
   50 max=max+nabs
   60 continue
      return
      end
c
c
      subroutine CAYSTAR(i,j,iw,k,n,m,max)
      integer i(k),j(k),iw(k)
      n2=0
      n3=0
      do 20 m1=1,k
      iw(m1)=i(m1)
      do 10 m2=1,k
      if (iw(m1) .ne. j(m2)) go to 5
      n3=n3+1
      iw(m1)=m2
      go to 20
    5 if (m2 .ne. k) go to 10
      iw(m1)=k+1
   10 continue
   20 continue
      do 50 l2=1,k
      l1=l2
      do 45 n1=1,k
      i1=iw(l1)
      if (i1 .eq. 0) go to 50
      if (i1 .ne. l2 .and. i1 .le. k) go to 40
      n2=n2+1
      iw(l1)=0
      go to 50
   40 iw(l1)=0
   45 l1=i1
   50 continue
      l=k-n3
      m=k-n2+1
      max=k
      return
      end
c
c
      subroutine HAMSTAR(i,j,k,n,m,max)
      integer i(k),j(k)
      n1=0
      n2=0
```

```
      do 10 m1=1,k
      if (i(m1) .ne. j(m1)) n1=n1+1
      do 10 m2=1,k
      if (i(m1) .ne. j(m2)) go to 10
      n2=n2+1
   10 continue
      l=k-n2
      m=n1+1
      max=2*k
      if (max .gt. n) max=n
      return
      end
C
C
      subroutine ULAMSTAR(i,j,iw,jw,k,n,m,max)
      integer i(k),j(k),iw(n),jw(n)
      m=0
      do 80 num=1,2
      do 5 l=1,n
    5 iw(l)=0
      do 10 l2=1,k
      do 10 l1=1,k
      if (j(l2) .ne. i(l1)) go to 10
      if (num .eq. 1) iw(l2)=l1
      if (num .eq. 2) iw(l1)=l2
   10 continue
      n1=n+1
      do 30 l1=1,k
      do 20 l2=1,k
   20 if (iw(l2) .eq. l1) go to 30
      n1=n1-1
      iw(n1)=l1
   30 continue
      n1=k
      do 70 l1=1,n
      if (iw(l1) .ne. 0) go to 70
      n1=n1+1
      iw(l1)=n1
   70 continue
      call ULAM(iw,jw,n,m1)
   80 if (m1 .gt. m) m=m1
      max=k
      return
      end
C
      subroutine ULAM(k,l,n,m)
      integer k(n),l(n)
      do 10 j=1,n
   10 l(j)=j
      do 30 i=1,n
      min1=i+1
      if (k(i) .eq. 1) min1=i-1
      if (l(1)+1 .lt. min1) min1=l(1)+1
```

```
      do 20 j=2,n
      min2=l(j)+1
      if (k(i) .eq. j .and. l(j-1) .lt. min2) min2=l(j-1)
      if (min1+1 .lt. min2) min2=min1+1
      l(j-1)=min1
   20 min1=min2
   30 l(n)=min1
      m=l(n)/2
      return
      end
```

2. The Subroutine GMETRIC

This FORTRAN subroutine (pages 152-157) calculates the distance between two partial rankings of n items, both of type $k_1,\ldots,k_{kn}$, with respect to a specified metric on S_n/S, where $S = S_{k_1} \times\ldots\times S_{k_{kn}}$. The metrics implemented here are the Hausdorff and fixed vector metrics of Sections IV.B and IV.C.

The subroutine is labeled GMETRIC, in recognition of the fact that a more general situation is considered than in the subroutine METRIC.

Input Parameters: The input parameters include n, kn, and the kn-vector k, discussed above. In addition, the input n-vectors i and j represent the two partial rankings. The vector i lists the n items in any order which is consistent with the first judge's partial ranking, while the vector j does the same for the second judge. Formally, $i = (\pi^{-1}(1),\ldots, \pi^{-1}(n))$ and $j = (\sigma^{-1}(1),\ldots, \sigma^{-1}(n))$, where $\pi,\sigma\varepsilon S_n$ are any elements of the cosets corresponding to the two partial rankings.

Just as in the subroutine METRIC, the input parameter metnum determines which metric on S_n/S is to be computed, by specifying which subroutine of GMETRIC is to be called. This is done in accordance with Table 5.

Output Parameters: The output parameter m is the value of the metric (except that R_{fv} and R^* are replaced by $2R_{fv}^2$ and R^{*2}, just as in METRIC).

The output parameter max is an upper bound on the maximum possible value of the metric:

Table 5. Subroutines of GMETRIC, the Associated Metrics on S_n/S, and the Corresponding Values of the Input Parameter metnum.

Subroutine	Metric	Value of metnum
GTAUSTAR	T^*	1
GRHOSTAR	R^*	2
GRHOFV	R_{fv}	3
GFOOTSTAR	F^*	4
GFOOTFV	F_{fv}	5
GCAYSTAR	C^*	6
GHAMSTAR	H^*	7
GULAMSTAR	U^*	8

$$\max \geq \max\{d^*(\pi^*,\sigma^*): \pi^*,\sigma^* \varepsilon S_n/S\}.$$

Again, R_{fv} and R^* are replaced by $2R_{fv}^2$ and R^{*2}.

The output matrix km gives the values of the n_{ij} (in the notation of Section IV.B). Thus, km(i,j) is the number of items placed in the i-th category by the first judge, and in the j-th category by the second.

Algorithms Used: The metrics of Section IV.B can all be expressed in terms of the n_{ij}. Therefore, the values of n_{ij} are computed in the initial part of the subroutine GMETRIC, and stored in km. The appropriate subroutine (GTAUSTAR, GRHOSTAR, etc.) is then called.

For most of the metrics, the rest of the computation is straightforward. Just as in the subroutine METRIC, the biggest challenge is to compute U^* efficiently. By Lemmas 1 and 4 of Section IV.D, this again boils down to computing efficiently the ordinary Ulam's metric on S_n. Again, Sellers' algorithm [S1] can be utilized (see page 141 for details).

```
C      THE SUBROUTINE GMETRIC
C
C
C      THIS SUBROUTINE CALCULATES THE DISTANCE BETWEEN
C       TWO PARTIAL RANKINGS WITH RESPECT TO A SPECIFIED
C       METRIC ON Sn / (Sk(1)X...XSk(kn)).
C
       subroutine GMETRIC(metnum,i,j,iw,jw,k,n,kn,km,m,max)
       integer i(n),j(n),iw(n),jw(n),k(kn),km(kn,kn)
       do 10 n1=1,kn
       do 10 n2=1,kn
    10 km(n1,n2)=0
       n3=0
       do 40 n1=1,kn
       n4=0
       do 30 n2=1,kn
       do 20 n5=n3+1,n3+k(n1)
       do 20 n6=n4+1,n4+k(n2)
    20 if (i(n5) .eq.  j(n6)) km(n1,n2)=km(n1,n2)+1
    30 n4=n4+k(n2)
    40 n3=n3+k(n1)
       l=metnum
       if (l .eq.  1) call GTAUSTAR(k,n,kn,km,m,max)
       if (l .eq.  2) call GRHOSTAR(k,n,kn,km,m,max)
       if (l .eq.  3) call GRHOFV(i,j,iw,jw,k,n,kn,m,max)
       if (l .eq.  4) call GFOOTSTAR(k,n,kn,km,m,max)
       if (l .eq.  5) call GFOOTFV(i,j,iw,jw,k,n,kn,m,max)
       if (l .eq.  6) call GCAYSTAR(k,n,kn,km,m,max)
       if (l .eq.  7) call GHAMSTAR(k,n,kn,km,m,max)
       if (l .eq.  8) call GULAMSTAR(iw,jw,k,n,kn,km,m,max)
       return
       end
C
C
       subroutine GTAUSTAR(k,n,kn,km,m,max)
       integer k(kn),km(kn,kn)
       max=(n*(n-1))/2
       n1=0
       n2=0
       n3=0
       do 10 i=1,kn-1
       do 10 ip=i+1,kn
       do 20 j=1,kn
    20 n2=n2+km(i,j)*km(ip,j)
       do 10 j=2,kn
       do 10 jp=1,j-1
    10 n1=n1+km(i,j)*km(ip,jp)
       do 30 i=1,kn
       do 30 j=2,kn
       do 30 jp=1,j-1
    30 n3=n3+km(i,j)*km(i,jp)
       if (n3 .gt.  n2) n2=n3
       m=n1+n2
```

```
      return
      end
C
C
      subroutine GRHOSTAR(k,n,kn,km,m,max)
      integer k(kn),km(kn,kn)
      n1=0
      n2=0
      do 10 i=1,kn-1
      do 10 j=i+1,kn
      n3=0
      do 11 l=1,j-1
   11 n3=n3-km(i,l)
      do 12 l=1,i
   12 n3=n3-km(l,j)
      n4=-k(i)-k(j)-n3
      do 13 l=i,j
      n3=n3+k(l)
   13 n4=n4+k(l)
      n1=n1+km(i,j)*n3*n3
   10 n2=n2+km(i,j)*n4*n4
      do 20 i=2,kn
      do 20 j=1,i-1
      n3=0
      do 21 l=i,kn
   21 n3=n3-km(l,j)
      do 22 l=j+1,kn
   22 n3=n3-km(i,l)
      n4=-k(j)-k(i)-n3
      do 23 l=j,i
      n3=n3+k(l)
   23 n4=n4+k(l)
      n1=n1+km(i,j)*n3*n3
   20 n2=n2+km(i,j)*n4*n4
      do 30 i=1,kn
      n3=0
      do 31 l=1,i
   31 n3=n3-km(i,l)-km(l,i)
      n3=n3+k(i)+km(i,i)
      n1=n1+km(i,i)*n3*n3
   30 n2=n2+km(i,i)*n3*n3
      if (n2 .gt. n1) n1=n2
      m=n1
      max=(n*(n-1)*(n+1))/3
      return
      end
C
C
      subroutine GRHOFV(i,j,iw,jw,k,n,kn,m,max)
      integer i(n),j(n),iw(n),jw(n),k(kn)
      max=(n*(n-1)*(n+1))/3
      m=0
      n1=1
```

```
      n2=0
      do 20 m1=1,kn
      n2=n2+k(m1)
      do 10 m2=n1,n2
      do 10 m3=1,n
      if (i(m2) .eq. m3) iw(m3)=n1+n2
   10 if (j(m2) .eq. m3) jw(m3)=n1+n2
   20 n1=n1+k(m1)
      do 30 n1=1,n
      n2=iw(n1)-jw(n1)
   30 m=m+n2*n2
      m=m/2
      return
      end
C
C
      subroutine GFOOTSTAR(k,n,kn,km,m,max)
      integer k(kn),km(kn,kn)
      max=2*(n-n/2)*(n/2)
      n1=0
      n2=0
      n3=0
      do 10 i=1,kn-1
      do 10 j=i+1,kn
      n4=0
      do 11 l=i,j
   11 n4=n4+k(l)
      n1=n1+km(i,j)*n4
      n4=0
      do 12 l=1,j-1
   12 n4=n4+km(i,l)
      n2=n2+km(i,j)*n4
      n4=0
      do 13 l=1,i
   13 n4=n4+km(l,j)
      n2=n2+km(i,j)*n4
      n4=0
      do 14 l=j,kn
   14 n4=n4+km(i,l)
      n3=n3+km(i,j)*n4
      n4=0
      do 15 l=i+1,kn
   15 n4=n4+km(l,j)
      n3=n3+km(i,j)*n4
   10 continue
      do 20 i=2,kn
      do 20 j=1,i-1
      n4=0
      do 21 l=j,i
   21 n4=n4+k(l)
      n1=n1+km(i,j)*n4
      n4=0
      do 22 l=i,kn
```

```
   22 n4=n4+km(l,j)
      n2=n2+km(i,j)*n4
      n4=0
      do 23 l=j+1,kn
   23 n4=n4+km(i,l)
      n2=n2+km(i,j)*n4
      n4=0
      do 24 l=1,i-1
   24 n4=n4+km(l,j)
      n3=n3+km(i,j)*n4
      n4=0
      do 25 l=1,j
   25 n4=n4+km(i,l)
      n3=n3+km(i,j)*n4
   20 continue
      do 30 i=1,kn
      n4=0
      do 31 l=1,i
   31 n4=n4+km(i,l)+km(l,i)
      n5=abs(float(k(i)+km(i,i)-n4))+.1
      n1=n1+km(i,i)*n5
   30 continue
      if (n3 .lt. n2) n2=n3
      m=n1-n2
      return
      end
C
C
      subroutine GFOOTFV(i,j,iw,jw,k,n,kn,m,max)
      integer i(n),j(n),iw(n),jw(n),k(kn)
      max=2*(m-m/2)*(m/2)
      m=0
      n1=1
      n2=0
      do 20 m1=1,kn
      n2=n2+k(m1)
      do 10 m2=n1,n2
      do 10 m3=1,n
      if (i(m2) .eq. m3) iw(m3)=n1+n2
   10 if (j(m2) .eq. m3) jw(m3)=n1+n2
   20 n1=n1+k(m1)
      do 30 n1=1,n
      nabs=iw(n1)-jw(n1)
      if (nabs .lt. 0) nabs=-nabs
   30 m=m+nabs
      m=m/2
      return
      end
C
C
      subroutine GCAYSTAR(k,n,kn,km,m,max)
      integer k(kn),km(kn,kn)
      max=n-1
```

```
      if (kn .ne. 2) go to 10
      m=km(1,2)
      go to 99
   10 if (kn .ne. 3) go to 90
      n1=0
      n2=0
      do 20 i=1,kn
      n1=n1+km(i,i)
      if (i .eq. kn) go to 20
      do 20 j=i+1,kn
      n3=km(i,j)
      if (km(j,i) .lt. n3) n3=km(j,i)
      n2=n2+n3
   20 continue
      n3=abs(float(km(1,2)-km(2,1)))+.1
      m=n-(n1+n2+n3)
      go to 99
   90 write (7,100)
  100 format(1x,'cayley-star cannot be computed')
      read (7,110) pause
  110 format(1x,f2.1)
   99 continue
      return
      end
C
C
      subroutine GHAMSTAR(k,n,kn,km,m,max)
      integer k(kn),km(kn,kn)
      n1=0
      do 10 i=1,kn
   10 n1=n1+km(i,i)
      m=n-n1
      n1=0
      do 20 n2=1,kn
   20 if (k(n2) .gt. n1) n1=k(n2)
      max=n
      if (n1 .gt. n/2) max=2*(n-n1)
      return
      end
C
C
      subroutine GULAMSTAR(iw,jw,k,n,kn,km,m,max)
      integer iw(n),jw(n),k(kn),km(kn,kn)
      max=n-1
      m=0
      do 20 num=1,2
      jw(1)=1
      do 5 n1=2,kn
    5 jw(n1)=jw(n1-1)+k(n1-1)
      n3=1
      do 10 n1=1,kn
      do 10 n2=kn,1,-1
      if (km(n2,n1) .eq. 0) go to 10
```

```
      do 10 n4=1,km(n2,n1)
      iw(n3)=jw(n2)
      n3=n3+1
      jw(n2)=jw(n2)+1
   10 continue
      call ULAM(iw,jw,n,m1)
      if (m1 .gt. m) m=m1
      do 20 n1=1,kn-1
      do 20 n2=n1+1,kn
      n3=km(n1,n2)
      km(n1,n2)=km(n2,n1)
   20 km(n2,n1)=n3
      return
      end
c
      subroutine ULAM(k,l,n,m)
      integer k(n),l(n)
      do 10 j=1,n
   10 l(j)=j
      do 30 i=1,n
      min1=i+1
      if (k(i) .eq. 1) min1=i-1
      if (l(1)+1 .lt. min1) min1=l(1)+1
      do 20 j=2,n
      min2=l(j)+1
      if (k(i) .eq. j .and. l(j-1) .lt. min2) min2=l(j-1)
      if (min1+1 .lt. min2) min2=min1+1
      l(j-1)=min1
   20 min1=min2
   30 l(n)=min1
      m=l(n)/2
      return
      end
```

APPENDIX C - FORTRAN SUBROUTINES FOR FITTING MALLOWS' MODEL TO PARTIALLY RANKED DATA

This appendix presents several FORTRAN subroutines for fitting Mallows' model to partially ranked data in S_n/S_{n-k}. Thus, the data is of the type where each judge lists his k favorite items out of n.

As explained in Section VI.A, Mallows' model has the form

$$\mathrm{Prob}(\pi^*) = Ce^{-\lambda d^*(\pi,\pi_0^*)} \qquad \forall\pi^* \varepsilon S_n/S_{n-k},$$

where $\pi_0^* \varepsilon S_n/S_{n-k}$ is a location parameter, $\lambda \geq 0$ is a dispersion parameter, d^* is a pre-specified metric on S_n/S_{n-k}, and $C = C(\lambda)$ is a "constant of proportionality." This parametrization has the advantage of displaying the model as an exponential family. However, for the purpose of fitting the model numerically, it will be convenient to use the slightly different parametrization

$$\mathrm{Prob}(\pi^*) = C\theta^{d^*(\pi^*,\pi_0^*)} \qquad \forall\pi^* \varepsilon S_n/S_{n-k},$$

where $\theta = e^{-\lambda} \varepsilon [0,1]$ replaces λ as the dispersion parameter.

The subroutines PINOUGHT, THETA, and METMEAN fit the parameters π_0^*, θ, and C by maximum likelihood.

1. The Subroutine PINOUGHT

This subroutine (pages 162-163) finds the maximum likelihood estimate $\hat{\pi}_0^*$ of π_0^*.

Input Parameters:

n,k: described above.

l: the number of distinct partial rankings $\pi_1^*,\ldots,\pi_l^*$ in the data set.

i: an $l \times k$ matrix, which encodes the distinct partial rankings $\pi_1^*,\ldots,\pi_l^*$. Specifically, the r-th row of i is $\pi_r^{-1}(1),\ldots,\pi_r^{-1}(k)$, where π_r is any full ranking which is consistent with the r-th partial ranking π_r^*. Thus, the r-th row of i lists in order the k favorite items, according to the r-th partial ranking π_r^*.

m: an l-vector, whose r-th component tells how many times the r-th distinct partial ranking π_r^* occurs in the data set.

metnum: an integer between 1 and 8, that specifies which metric d^* is to be used in the model, in accordance with Table 1 of Appendix B. (Note: If metnum = 3, the metric used is actually $2R_{fv}$ instead of R_{fv}).

jb: an n-vector, input interactively, which stores an initial guess of the central partial ranking $\pi_0^* \varepsilon S_n/S_{n-k}$. More precisely, we make an initial guess of π_0^*, and then feed into jb the values $\pi^{-1}(1),\ldots,\pi^{-1}(n)$, where π is any full ranking consistent with our guessed partial ranking.

iw,jw: n-vectors, which provide necessary storage space for calculating some of the metrics. Their input values need not be specified.

j,il: n-vectors, which provide necessary storage space for computing the maximum likelihood estimate $\hat{\pi}_0^*$. Their input values need not be specified.

Output Parameters:

jb: an n-vector, which stores the maximum likelihood estimate $\hat{\pi}_0^*$. More precisely, it lists $\pi^{-1}(1),\ldots,\pi^{-1}(n)$, where π is a full ranking consistent with $\hat{\pi}_0^*$.

sb: the sum of the distances of all partial rankings in the data set from the maximum likelihood estimate $\hat{\pi}_0^*$. Thus

$$sb = \sum_{\pi^* \in \text{ data set}} d^*(\pi^*,\hat{\pi}_0^*),$$

with "repeats counted": if a partial ranking occurs more than once in the data set, it occurs more than once in the sum.

sd: the total number of judges — i.e., the total number of partial rankings in the data set, with repeats counted.

max: the maximum possible value of the metric d^*:

$$\max = \max \{ d^*(\pi^*,\sigma^*) : \pi^*,\sigma^* \varepsilon S_n/S_{n-k} \}.$$

The Algorithm: As derived in Section VI.A.4, the maximum likelihood estimate $\hat{\pi}_0^*$ is that partial ranking which minimizes the above sum

$$sb = \sum_{\pi^* \in \text{ data set}} d^*(\pi^*,\hat{\pi}_0^*).$$

The algorithm for finding $\hat{\pi}_0^*$ is to make an initial guess, calculate sb for this guess, and then to modify the guess by successive pairwise adjacent transpositions, until a smaller value of sb is obtained. The partial ranking which yielded this smaller value of sb then becomes our <u>new</u> guess for the maximum likelihood estimate $\hat{\pi}_0^*$.

The algorithm continues in this way, until it hits upon a value of $\hat{\pi}_0^*$ which minimizes the sum sb.

```
C      THE SUBROUTINE PINOUGHT
C
C
C      THIS SUBROUTINE FITS THE LOCATION PARAMETER PI NOUGHT
C       BY MAXIMUM LIKELIHOOD.
C
       subroutine PINOUGHT(metnum,i,m,iw,jw,l,n,k,j,i1,jb,
      X   sb,sd,max)
       integer i(l,k),m(l),j(n),jb(n),i1(k),iw(n),jw(n)
       sb=10.**10.
       n4=0
       n7=0
       sd=0
       do 5 n1=1,l
     5 sd=sd+float(m(n1))
       write(7,102)sd
   102 format(1x,'number of judges =',f10.0)
       write(7,100)
   100 format(1x,'INPUT INITIAL jb(1),...,jb(n)')
       write(7,101)
   101 format(1x,'--format(1x,i3)')
       do 10 n1=1,n
       read(7,110)jb(n1)
   110 format(1x,i3)
    10 j(n1)=jb(n1)
    15 do 80 n5=1,n
       s=0
       do 20 n1=1,l
       do 30 n2=1,k
    30 i1(n2)=i(n1,n2)
       call METRIC(metnum,i1,j,iw,jw,k,n,n3,max)
       if (metnum .ne. 2 .and. metnum .ne. 3)
      X   s=s+float(m(n1)*n3)/sd
       if (metnum .eq. 2 .or. metnum .eq. 3)
      X   s=s+float(m(n1))*sqrt(float(n3))/sd
    20 continue
       sbound=sb-.001
       if (s .ge. sbound) go to 41
       n4=0
       sb=s
       do 40 n1=1,n
    40 jb(n1)=j(n1)
       n7=n7+1
       go to 82
    41 n4=n4+1
       if (n4 .eq. n) go to 99
       if (n5 .eq. 1) go to 45
       if (n5-1 .ge. k) go to 42
       n6=j(n5)
       j(n5)=j(n5-1)
       n5m1=n5-1
       j(n5m1)=n6
       go to 45
```

```
 42 n6=j(n5)
    j(n5)=j(k)
    j(k)=n6
 45 if (n5 .ge. k) go to 50
    n6=j(n5)
    j(n5)=j(n5+1)
    n5p1=n5+1
    j(n5p1)=n6
    n7=n7+1
    go to 80
 50 if (n5 .eq. n) go to 80
    n6=j(n5+1)
    n5p1=n5+1
    j(n5p1)=j(k)
    j(k)=n6
    n7=n7+1
 80 continue
 82 if (n7 .le. 20) go to 15
 85 write(7,140)
140 format(1x,'TOO MANY REPETITIONS')
 99 continue
    return
    end
```

2. The Subroutines THETA and METMEAN

Once the maximum likelihood estimate $\hat{\pi}_0^*$ has been determined by the subroutine PINOUGHT, the FORTRAN subroutines THETA (page 166) and METMEAN (page 167) find the maximum likelihood estimate of the dispersion parameter θ, as well as C, the constant of proportionality in Mallows' model.

The subroutine METMEAN is subsidiary to the subroutine THETA. Hence, input and output parameters are described below for THETA only.

Input Parameters for the Subroutine THETA

n,k,metnum,iw,jw: These input parameters have the same meaning as in the subroutine PINOUGHT.

sm: the sample mean of the distances $d^*(\cdot,\hat{\pi}_0^*)$, obtained by averaging the values of $d^*(\cdot,\hat{\pi}_0^*)$ over all the partial rankings in the data set. Thus, the value of sm is the quotient sb/sd, where sb and sd are output parameters from the subroutine PINOUGHT.

i2,j2: k-vectors, which provide necessary storage space for computing the _theoretical_ mean of d^*. The input values of i2 and j2 need not be specified.

Output Parameters for the Subroutine THETA:

t: stores $\hat{\theta}$, the maximum likelihood estimate of the dispersion parameter θ.

s2: stores the sum $\sum_{\pi^* \varepsilon S_n/S_{n-k}} \hat{\theta}^{d^*(\pi^*,\hat{\pi}_0^*)}$. The constant of proportionality C is easily obtained from s2, since $C = (s2)^{-1}$.

```
C      THE SUBROUTINE THETA
C
C
C      THIS SUBROUTINE FITS THE DISPERSION PARAMETER
C       THETA BY MAXIMUM LIKELIHOOD.
C
       subroutine THETA(metnum,i2,j2,iw,jw,n,k,sm,t,s2)
       integer i2(k),j2(k),iw(n),jw(n)
       call METMEAN(metnum,1.0,i2,j2,iw,jw,n,k,e,s2)
       if (sm .le. e) go to 25
       t=1.0
       go to 30
    25 tu=1.0
       tl=0
       do 30 l=1,15
       t=(tu+tl)/2.
       call METMEAN(metnum,t,i2,j2,iw,jw,n,k,e,s2)
       if (e .gt. sm) tu=t
       if (e .le. sm) tl=t
    30 continue
       return
       end
```

```
C     THE SUBROUTINE METMEAN
C
C
C     THIS SUBROUTINE CALCULATES THE EXPECTED VALUE
C      OF THE METRIC, FOR A SPECIFIED VALUE OF
C      THE DISPERSION PARAMETER THETA.
C
      subroutine METMEAN(metnum,t,i,j,iw,jw,n,k,e,s2)
      integer i(k),j(k),iw(n),jw(n)
      s1=0
      s2=0
      do 10 m=1,k
   10 i(m)=m
      m1=n**k
      do 30 m=1,m1
      m2=m-1
      do 20 m3=1,k
      j(m3)=1+m2/(n**(k-m3))
      if (m3 .eq. 1) go to 15
      do 15 m4=1,m3-1
      if (j(m3) .eq. j(m4)) go to 30
   15 continue
   20 m2=m2-(j(m3)-1)*(n**(k-m3))
      call METRIC(metnum,i,j,iw,jw,k,n,n1,max)
      x=float(n1)
      if (metnum .eq. 2 .or. metnum .eq. 3) x=sqrt(x)
      if (t .ne. 0) go to 25
      if (x .ne. 0) t1=0
      if (x .eq. 0) t1=1
      go to 26
   25 t1=t**x
   26 s1=s1+x*t1
      s2=s2+t1
   30 continue
      e=s1/s2
      return
      end
```

The Algorithm: By Section VI.A.4, the maximum likelihood estimate $\hat{\theta}$ is that value of θ which makes sm, the observed mean distance from $\hat{\pi}_0^*$, agree with the expected distance $E_{\hat{\pi}_0^*,\theta}[d^*(\cdot,\hat{\pi}_0^*)]$. Here $E_{\hat{\pi}_0^*,\theta}$ denotes taking the expectation with respect to the Mallows model with parameters $\hat{\pi}_0^*$ and θ. Note that $E_{\hat{\pi}_0^*,\theta}[d^*(\cdot,\hat{\pi}_0^*)]$ is an increasing function of θ.

Initially $\hat{\theta}$ is known to lie in the interval [0,1], so it seems reasonable to first try the value $\theta = \frac{1}{2}$. The subroutine METMEAN calculates $E_{\hat{\pi}_0^*,\theta}[d^*(\cdot,\hat{\pi}_0^*)]$ for any desired value of θ, and stores it in the output parameter e. Therefore, this subroutine is called, and the expectation is computed initially for $\theta = \frac{1}{2}$. If the expectation is greater than the sample mean sm, then $\hat{\theta}$ must lie in the interval $[0,\frac{1}{2}]$; otherwise, $\hat{\theta}$ lies in $[\frac{1}{2},1]$. In either case, $\hat{\theta}$ is known to lie in a new interval, whose width is half that of the initial interval [0,1]. The process now iterates, taking the midpoint of this new interval as the trial value of θ, and once again calling the subroutine METMEAN. After fifteen iterations, $\hat{\theta}$ is known to lie in an interval of width 2^{-15}; this is considered an acceptable margin of error.

APPENDIX D - TABLES OF THE DISTRIBUTIONS OF THE METRICS ON S_n/S_{n-k}

This appendix presents tables of the distributions of the various metrics T^*, R^*, R_{fv}, F^*, F_{fv}, C^*, H^*, and U^* on the coset space S_n/S_{n-k}. Thus, we consider partially ranked data of the form where each judge specifies his first through k-th choices out of n items. The tables encode the probability distribution of the distance $d^*(\pi^*,\sigma^*)$ between two such partial rankings $\pi^*,\sigma^* \varepsilon S_n/S_{n-k}$, under the assumption that π^* and σ^* are generated, independently, from a uniform distribution on all possible partial rankings.

The preceding distribution depends on three things: the metric d^*, the value of k, and the value of n. In this appendix, each table fixes a particular metric d^* and a particular value of k, and then encodes the distributions for varying values of n. This is done in accordance with Table 6.

Since Table 6 summarizes the differences <u>among</u> Tables 7 through 38, it remains to describe the underlying format <u>within</u> each of these tables. As already noted, each table contains several distributions, for different values of n; these values of n are listed in the first column of each table. (Of course, when n=k or when n=k+1, the induced metric on S_n/S_{n-k} coincides with the ordinary metric on the full permutation group S_n.)

The second column of each table — for example, the column labelled "T^*" in Table 7 — lists the possible values of the metric.

Table 6. Guide to Tables 7-38.

Table	Metric	k	Page	Table	Metric	k	Page
7	T^*	2	171	23	F_{fv}	2	189
8	T^*	3	172	24	F_{fv}	3	190
9	T^*	4	173	25	F_{fv}	4	191
10	T^*	5	174	26	F_{fv}	5	192
11	R^*	2	175	27	C^*	2	193
12	R^*	3	176	28	C^*	3	194
13	R^*	4	177	29	C^*	4	195
14	R^*	5	178	30	C^*	5	196
15	R_{fv}	2	180	31	H^*	2	197
16	R_{fv}	3	181	32	H^*	3	198
17	R_{fv}	4	182	33	H^*	4	199
18	R_{fv}	5	184	34	H^*	5	200
19	F^*	2	185	35	U^*	2	201
20	F^*	3	186	36	U^*	3	202
21	F^*	4	187	37	U^*	4	203
22	F^*	5	188	38	U^*	5	204

Explanation: Each table fixes a particular metric and a particular value of k, and then encodes the distributions of that metric on S_n/S_{n-k}, for varying values of n.

The only exceptions to this are for the metrics R^* and R_{fv}, where the second column lists the possible values of R^{*2} and $2R_{fv}^2$, respectively. (These transformations are performed because R^{*2} and $2R_{fv}^2$ are always integer-valued, whereas R^* and R_{fv} are not. Throughout the rest of this description, it will be implicitly understood that "d^*" stands for R^{*2} or $2R_{fv}^2$, rather than R^* or R_{fv}.)

The third column, labeled "Frequency," counts the number of partial rankings σ^* which lie at a given distance from any fixed partial ranking π^*. Hence, the entry in column 3 is

$$\#\{\sigma^* \varepsilon S_n/S_{n-k}: d^*(\pi^*,\sigma^*) = m\},$$

where m is the entry in column 2. To find the <u>probability</u> that the distance m is attained, we need only divide the entry in the Frequency column by the k-fold product $n_{(k)} = n\cdot(n-1)\cdot\ldots\cdot(n-k+1)$.

The fourth column contains the <u>cumulative probability</u> that the value of the metric is less than or equal to a given value. In other words, the entry in column 4 is

$$\text{Prob}\{\sigma^* \varepsilon S_n/S_{n-k}: d^*(\pi^*,\sigma^*) \leq m\},$$

where m is the entry in column 2, and σ^* is assumed to be uniformly distributed over all possible partial rankings. Of course, if two partial rankings are a distance m apart, the entry in column 4 can be interpreted as a p-value for these rankings, measuring the extent to which they are "significantly" correlated.

Table 7. Distributions of T^* on S_n/S_{n-k}, for k = 2.

	T^*	Frequency	Cumulative Probability
n=2:	0	1	.5000
	1	1	1.0000
n=3:	0	1	.1667
	1	2	.5000
	2	2	.8333
	3	1	1.0000
n=4:	0	1	.0833
	1	1	.1667
	2	2	.3333
	3	4	.6667
	4	2	.8333
	5	2	1.0000
n=5:	0	1	.0500
	1	1	.1000
	3	3	.2500
	4	6	.5500
	5	3	.7000
	7	6	1.0000
n=6:	0	1	.0333
	1	1	.0667
	4	4	.2000
	5	8	.4667
	6	4	.6000
	9	12	1.0000
n=7:	0	1	.0238
	1	1	.0476
	5	5	.1667
	6	10	.4048
	7	5	.5238
	11	20	1.0000
n=8:	0	1	.0179
	1	1	.0357
	6	6	.1429
	7	12	.3571
	8	6	.4643
	13	30	1.0000
n=9:	0	1	.0139
	1	1	.0278
	7	7	.1250
	8	14	.3194
	9	7	.4167
	15	42	1.0000
n=10:	0	1	.0111
	1	1	.0222
	8	8	.1111
	9	16	.2889
	10	8	.3778
	17	56	1.0000
n=11:	0	1	.0091
	1	1	.0182
	9	9	.1000
	10	18	.2636
	11	9	.3455
	19	72	1.0000
n=12:	0	1	.0076
	1	1	.0152
	10	10	.0909
	11	20	.2424
	12	10	.3182
	21	90	1.0000
n=13:	0	1	.0064
	1	1	.0128
	11	11	.0833
	12	22	.2244
	13	11	.2949
	23	110	1.0000
n=14:	0	1	.0055
	1	1	.0110
	12	12	.0769
	13	24	.2088
	14	12	.2747
	25	132	1.0000
n=15:	0	1	.0048
	1	1	.0095
	13	13	.0714
	14	26	.1952
	15	13	.2571
	27	156	1.0000

Table 8. Distributions of T^* on S_n/S_{n-k}, for k = 3.

	T^*	Frequency	Cumulative Probability
n=3:	0	1	.1667
	1	2	.5000
	2	2	.8333
	3	1	1.0000
n=4:	0	1	.0417
	1	3	.1667
	2	5	.3750
	3	6	.6250
	4	5	.8333
	5	3	.9583
	6	1	1.0000
n=5:	0	1	.0167
	1	2	.0500
	2	4	.1167
	3	7	.2333
	4	10	.4000
	5	12	.6000
	6	10	.7667
	7	8	.9000
	8	4	.9667
	9	2	1.0000
n=6:	0	1	.0083
	1	2	.0250
	2	2	.0417
	3	4	.0750
	4	9	.1500
	5	15	.2750
	6	15	.4000
	7	15	.5250
	8	15	.6500
	9	18	.8000
	10	12	.9000
	11	6	.9500
	12	6	1.0000
n=7:	0	1	.0048
	1	2	.0143
	2	2	.0238
	3	1	.0286
	4	4	.0476
	5	12	.1048
	6	20	.2000
	7	20	.2952
	8	12	.3524
	9	16	.4286
	10	24	.5429
	11	36	.7143
	12	24	.8286
	13	12	.8857
	15	24	1.0000
n=8:	0	1	.0030
	1	2	.0089
	2	2	.0149
	3	1	.0179
	5	5	.0327
	6	15	.0774
	7	25	.1518
	8	25	.2262
	9	15	.2708
	10	5	.2857
	11	20	.3452
	12	40	.4643
	13	60	.6429
	14	40	.7619
	15	20	.8214
	18	60	1.0000
n=9:	0	1	.0020
	1	2	.0060
	2	2	.0099
	3	1	.0119
	6	6	.0238
	7	18	.0595
	8	30	.1190
	9	30	.1786
	10	18	.2143
	11	6	.2262
	13	30	.2857
	14	60	.4048
	15	90	.5833
	16	60	.7024
	17	30	.7619
	21	120	1.0000

Table 9. Distributions of T^* on S_n/S_{n-k}, for k = 4.

	T^*	Frequency	Cumulative Probability
n=4:	0	1	.0417
	1	3	.1667
	2	5	.3750
	3	6	.6250
	4	5	.8333
	5	3	.9583
	6	1	1.0000
n=5:	0	1	.0083
	1	4	.0417
	2	9	.1167
	3	15	.2417
	4	20	.4083
	5	22	.5917
	6	20	.7583
	7	15	.8833
	8	9	.9583
	9	4	.9917
	10	1	1.0000
n=6:	0	1	.0028
	1	3	.0111
	2	7	.0306
	3	14	.0694
	4	23	.1333
	5	35	.2306
	6	45	.3556
	7	52	.5000
	8	52	.6444
	9	46	.7722
	10	36	.8722
	11	24	.9389
	12	14	.9778
	13	6	.9944
	14	2	1.0000
n=7:	0	1	.0012
	1	3	.0048
	2	5	.0107
	3	9	.0214
	4	17	.0417
	5	30	.0774
	6	46	.1321
	7	63	.2071
	8	75	.2964
	9	87	.4000
	10	93	.5107
	11	96	.6250
	12	93	.7357
	13	78	.8286
	14	60	.9000
	15	42	.9500
	16	24	.9786
	17	12	.9929
	18	6	1.0000
n=8:	0	1	.0006
	1	3	.0024
	2	5	.0054
	3	6	.0089
	4	9	.0143
	5	19	.0256
	6	37	.0476
	7	60	.0833
	8	76	.1286
	9	88	.1810
	10	96	.2381
	11	120	.3095
	12	148	.3916
	13	172	.5000
	14	168	.6000
	15	156	.6929
	16	132	.7714
	17	108	.8357
	18	108	.9000
	19	72	.9429
	20	48	.9714
	21	24	.9857
	22	24	1.0000

Table 10. Distributions of T^* on S_n/S_{n-k}, for k = 5.

	T^*	Frequency	Cumulative Probability		T^*	Frequency	Cumulative Probability
n=5:	0	1	.0083	n=7:	0	1	.0004
	1	4	.0417		1	4	.0020
	2	9	.1167		2	11	.0063
	3	15	.2417		3	25	.0163
	4	20	.4083		4	48	.0353
	5	22	.5917		5	82	.0679
	6	20	.7583		6	126	.1179
	7	15	.8833		7	177	.1881
	8	9	.9583		8	227	.2782
	9	4	.9917		9	268	.3845
	10	1	1.0000		10	291	.5000
					11	292	.6159
n=6:	0	1	.0014		12	270	.7230
	1	5	.0083		13	230	.8143
	2	14	.0278		14	180	.8857
	3	29	.0681		15	128	.9365
	4	49	.1361		16	82	.9690
	5	71	.2347		17	46	.9873
	6	90	.3597		18	22	.9960
	7	101	.5000		19	8	.9992
	8	101	.6403		20	2	1.0000
	9	90	.7653				
	10	71	.8639				
	11	49	.9319				
	12	29	.9722				
	13	14	.9917				
	14	5	.9986				
	15	1	1.0000				

Table 11. Distributions of R^* on S_n/S_{n-k}, for k = 2.

	R^{*2}	Frequency	Cumulative Probability
n=2:	0	1	.5000
	2	1	1.0000
n=3:	0	1	.1667
	2	2	.5000
	6	2	.8333
	8	1	1.0000
n=4:	0	1	.0833
	2	1	.1667
	6	2	.3333
	12	4	.6667
	14	2	.8333
	18	2	1.0000
n=5:	0	1	.0500
	2	1	.1000
	12	3	.2500
	20	6	.5500
	22	3	.7000
	32	6	1.0000
n=6:	0	1	.0333
	2	1	.0667
	20	4	.2000
	30	8	.4667
	32	4	.6000
	50	12	1.0000
n=7:	0	1	.0238
	2	1	.0476
	30	5	.1667
	42	10	.4048
	44	5	.5238
	72	20	1.0000
n=8:	0	1	.0179
	2	1	.0357
	42	6	.1429
	56	12	.3571
	58	6	.4643
	98	30	1.0000
n=9:	0	1	.0139
	2	1	.0278
	56	7	.1250
	72	14	.3194
	74	7	.4167
	128	42	1.0000
n=10:	0	1	.0111
	2	1	.0222
	72	8	.1111
	90	16	.2889
	92	8	.3778
	162	56	1.0000
n=11:	0	1	.0091
	2	1	.0182
	90	9	.1000
	110	18	.2636
	112	9	.3455
	200	72	1.0000
n=12:	0	1	.0076
	2	1	.0152
	110	10	.0909
	132	20	.2424
	143	10	.3182
	242	90	1.0000
n=13:	0	1	.0064
	2	1	.0128
	132	11	.0833
	156	22	.2244
	158	11	.2949
	288	110	1.0000
n=14:	0	1	.0055
	2	1	.0110
	156	12	.0769
	182	24	.2088
	184	12	.2747
	338	132	1.0000
n=15:	0	1	.0048
	2	1	.0095
	182	13	.0714
	210	26	.1952
	212	13	.2571
	392	156	1.0000

Table 12. Distributions of R^* on S_n/S_{n-k}, for k = 3.

	R^{*2}	Frequency	Cumulative Probability		R^{*2}	Frequency	Cumulative Probability
n=3:	0	1	.1667	n=6:	0	1	.0083
	2	2	.5000		2	2	.0250
	6	2	.8333		6	2	.0417
	8	1	1.0000		8	1	.0500
					12	3	.0750
n=4:	0	1	.0417		14	3	.1000
	2	3	.1667		20	6	.1500
	4	1	.2083		22	3	.1750
	6	4	.3750		24	6	.2250
	8	2	.4583		30	9	.3000
	10	2	.5417		32	18	.4500
	12	2	.6250		36	9	.5250
	14	4	.7917		38	3	.5500
	16	1	.8333		42	12	.6500
	18	3	.9583		46	6	.7000
	20	1	1.0000		50	12	.8000
					52	12	.9000
n=5:	0	1	.0167		56	6	.9500
	2	2	.0500		62	6	1.0000
	6	4	.1167				
	8	3	.1667	n=7:	0	1	.0048
	12	4	.2333		2	2	.0143
	14	2	.2667		6	2	.0238
	16	4	.3333		8	1	.0286
	18	2	.3667		20	4	.0476
	20	4	.4333		22	4	.0667
	22	10	.6000		30	8	.1048
	26	10	.7667		32	4	.1238
	28	2	.8000		34	8	.1619
	30	2	.8333		40	4	.1810
	32	4	.9000		42	8	.2190
	36	4	.9667		44	16	.2952
	38	2	1.0000		48	12	.3524
					50	16	.4286
					62	24	.5429
					66	12	.6000
					72	24	.7143
					74	24	.8286
					78	12	.8856
					92	24	1.0000

Table 13. Distributions of R^* on S_n/S_{n-k}, for k = 4.

	R^{*2}	Frequency	Cumulative Probability
n=4:	0	1	.0417
	2	3	.1667
	4	1	.2083
	6	4	.3750
	8	2	.4583
	10	2	.5417
	12	2	.6250
	14	4	.7917
	16	1	.8333
	18	3	.9583
	20	1	1.0000
n=5:	0	1	.0083
	2	4	.0417
	4	3	.0667
	6	6	.1167
	8	7	.1750
	10	6	.2250
	12	4	.2583
	14	10	.3417
	16	6	.3917
	18	10	.4750
	20	6	.5250
	22	10	.6083
	24	6	.6583
	26	10	.7417
	28	4	.7750
	30	6	.8250
	32	7	.8833
	34	6	.9333
	36	3	.9583
	38	4	.9917
	40	1	1.0000
n=6:	0	1	.0028
	2	3	.0111
	4	1	.0139
	6	6	.0306
	8	6	.0472
	10	2	.0528
	12	10	.0806
	14	12	.1139
	16	7	.1333
	18	5	.1472
	20	11	.1778
	22	14	.2167
	24	8	.2389
	26	14	.2778
	28	10	.3056
	30	14	.3444
	32	26	.4167
	34	6	.4333
	36	24	.5000
	38	24	.5667
	40	20	.6222
	42	10	.6500
	44	22	.7111
	46	14	.7500
	48	18	.8000
	50	6	.8167
	52	16	.8611
	54	6	.8778
	56	10	.9056
	58	4	.9167
	60	8	.9389
	62	14	.9778
	66	6	.9944
	68	2	1.0000

Table 14. Distributions of R^* on S_n/S_{n-k}, for k = 5.

	R^{*2}	Frequency	Cumulative Probability		R^{*2}	Frequency	Cumulative Probability
n=5	0	1	.0083	n=6:	0	1	.0014
	2	4	.0417		2	5	.0083
	4	3	.0667		4	6	.0167
	6	6	.1167		6	9	.0292
	8	7	.1750		8	16	.0514
	10	6	.2250		10	12	.0681
	12	4	.2583		12	14	.0875
	14	10	.3417		14	24	.1208
	16	6	.3917		16	20	.1486
	18	10	.4750		18	21	.1778
	20	6	.5250		20	23	.2097
	22	10	.6083		22	28	.2486
	24	6	.6583		24	24	.2819
	26	10	.7417		26	34	.3292
	28	4	.7750		28	20	.3569
	30	6	.8250		30	32	.4014
	31	7	.8833		32	42	.4597
	34	6	.9333		34	29	.5000
	36	3	.9583		36	29	.5403
	38	4	.9917		38	42	.5986
	40	1	1.0000		40	32	.6431
					42	20	.6708
					44	34	.7181
					46	24	.7514
					48	28	.7903
					50	23	.8222
					52	21	.8514
					54	20	.8792
					56	24	.9125
					58	14	.9319
					60	12	.9486
					62	16	.9708
					64	9	.9833
					66	6	.9917
					68	5	.9986
					70	1	1.0000

Table 14. Continued.

	R^{*2}	Frequency	Cumulative Probability
n=7:	0	1	.0004
	2	4	.0020
	4	3	.0032
	6	8	.0063
	8	13	.0115
	10	8	.0147
	12	16	.0210
	14	24	.0306
	16	18	.0377
	18	28	.0488
	20	34	.0623
	22	32	.0750
	24	38	.0901
	26	36	.1044
	28	32	.1171
	30	38	.1321
	32	63	.1571
	34	32	.1698
	36	67	.1964
	38	56	.2187
	40	65	.2444
	42	38	.2595
	44	80	.2913
	46	64	.3167
	48	98	.3556
	50	72	.3841
	52	78	.4151
	54	76	.4452
	56	98	.4841
	58	74	.5135
	60	106	.5556
	62	76	.5857
	64	96	.6238
	66	76	.6540
	68	96	.6921
	70	50	.7119
	72	80	.7437
	74	78	.7746
	76	52	.7952
	78	62	.8198
	80	70	.8476
	82	36	.8519
	84	42	.8786
	86	36	.8929
	88	46	.9111
	90	32	.9238
	92	42	.9405
	94	22	.9492
n=7:	96	36	.9635
	98	26	.9738
	100	20	.9817
	102	14	.9873
	104	18	.9944
	106	4	.9960
	108	8	.9992
	110	2	1.0000

Table 15. Distributions of R_{fv} on S_n/S_{n-k}, for k = 2.

	$2R_{fv}^2$	Frequency	Cumulative Probability
n=2:	0	1	.5000
	4	1	1.0000
n=3:	0	1	.1667
	4	2	.5000
	12	2	.8333
	16	1	1.0000
n=4:	0	1	.0833
	4	1	.1667
	9	2	.3333
	19	4	.6667
	25	2	.8333
	34	2	1.0000
n=5:	0	1	.0500
	4	1	.1000
	16	3	.2500
	28	6	.5500
	36	3	.7000
	52	6	1.0000
n=6:	0	1	.0333
	4	1	.0667
	25	4	.2000
	39	8	.4667
	49	4	.6000
	74	12	1.0000
n=7:	0	1	.0238
	4	1	.0476
	36	5	.1667
	52	10	.4048
	64	5	.5238
	100	20	1.0000
n=8:	0	1	.0179
	4	1	.0357
	49	6	.1429
	67	12	.3571
	81	6	.4643
	130	30	1.0000
n=9:	0	1	.0139
	4	1	.0278
	64	7	.1250
	84	14	.3194
	100	7	.4167
	164	42	1.0000
n=10:	0	1	.0111
	4	1	.0222
	81	8	.1111
	103	16	.2889
	121	8	.3778
	202	56	1.0000
n=11:	0	1	.0091
	4	1	.0182
	100	9	.1000
	124	18	.2636
	144	9	.3455
	244	72	1.0000
n=12:	0	1	.0076
	4	1	.0152
	121	10	.0909
	147	20	.2424
	169	10	.3182
	290	90	1.0000
n=13:	0	1	.0064
	4	1	.0128
	144	11	.0833
	172	22	.2244
	196	11	.2949
	340	110	1.0000
n=14:	0	1	.0055
	4	1	.0110
	169	12	.0769
	199	24	.2088
	225	12	.2747
	394	132	1.0000
n=15:	0	1	.0048
	4	1	.0095
	196	13	.0714
	228	26	.1952
	256	13	.2571
	452	156	1.0000

Table 16. Distributions of R_{fv} on S_n/S_{n-k}, for k = 3.

	$2R^2_{fv}$	Frequency	Cumulative Probability
n=3:	0	1	.1667
	4	2	.5000
	12	2	.8333
	16	1	1.0000
n=4:	0	1	.0417
	2	3	.1667
	4	1	.2083
	6	4	.3750
	8	2	.4583
	10	2	.5417
	12	2	.6250
	14	4	.7917
	16	1	.8333
	18	3	.9583
	20	1	1.0000
n=5:	0	1	.0167
	4	2	.0500
	9	2	.0833
	12	2	.1167
	13	2	.1500
	16	1	.1667
	19	4	.2333
	25	2	.2667
	27	4	.3333
	33	4	.4000
	34	2	.4333
	37	4	.5000
	39	4	.5667
	41	2	.6000
	47	4	.6667
	48	4	.7333
	49	2	.7667
	53	2	.8000
	58	2	.8333
	62	4	.9000
	68	4	.9667
	74	2	1.0000
n=6:	0	1	.0083
	4	2	.0250
	12	2	.0417
	16	4	.0750
	20	3	.1000
	28	6	.1500
	36	9	.2250
	44	6	.2750
	48	6	.3250
	52	15	.4500
	60	6	.5000
	64	3	.5250
	68	15	.6500
	80	6	.7000
	84	12	.8000
	92	12	.9000
	100	6	.9500
	116	6	1.0000
n=7:	0	1	.0048
	4	2	.0143
	12	2	.0238
	16	1	.0286
	25	4	.0476
	29	4	.0667
	39	8	.1048
	47	8	.1429
	49	4	.1619
	57	8	.2000
	61	8	.2381
	65	4	.2571
	67	8	.2952
	74	12	.3524
	75	8	.3905
	81	4	.4095
	85	4	.4286
	92	24	.5429
	106	12	.6000
	110	24	.7143
	120	24	.8286
	130	12	.8857
	155	24	1.0000

Table 17. Distributions of R_{fv} on S_n/S_{n-k}, for k = 4.

	$2R^2_{fv}$	Frequency	Cumulative Probability		$2R^2_{fv}$	Frequency	Cumulative Probability
n=4:	0	1	.0417	n=6:	0	1	.0028
	4	3	.1667		4	3	.0111
	8	1	.2083		8	1	.0139
	12	4	.3750		9	2	.0194
	16	2	.4583		12	4	.0306
	20	2	.5417		13	4	.0417
	24	2	.6250		16	2	.0472
	28	4	.7917		19	4	.0583
	32	1	.8333		20	2	.0639
	36	3	.9583		21	4	.0750
	40	1	1.0000		23	4	.0861
					24	2	.0917
n=5:	0	1	.0083		25	4	.1028
	4	4	.0417		27	4	.1139
	8	3	.0667		28	4	.1250
	12	6	.1167		29	2	.1306
	16	7	.1750		32	1	.1333
	20	6	.2250		33	4	.1444
	24	4	.2583		34	2	.1500
	28	10	.3417		35	4	.1611
	32	6	.3917		36	3	.1694
	36	10	.4750		37	4	.1806
	40	6	.5250		38	2	.1861
	44	10	.6083		39	8	.2083
	48	6	.6583		40	1	.2111
	52	10	.7417		41	6	.2278
	56	4	.7750		43	4	.2389
	60	6	.8250		47	8	.2611
	64	7	.8833		48	4	.2722
	68	6	.9333		49	6	.2889
	72	3	.9583		51	4	.3000
	76	4	.9917		53	10	.3278
	80	1	1.0000		55	8	.3500
					56	4	.3611
					57	8	.3833
					58	2	.3889
					59	4	.4000
					61	6	.4167
					62	4	.4278
					63	8	.4500
					65	6	.4667
					66	4	.4778
					67	8	.5000
					68	4	.5111
					70	4	.5222
					71	8	.5444
					73	4	.5556
					74	8	.5778

Table 17. Continued.

	$2R^2_{fv}$	Frequency	Cumulative Probability		$2R^2_{fv}$	Frequency	Cumulative Probability
n=6:	75	8	.6000	n=7:	0	1	.0012
	76	4	.6111		4	3	.0048
	77	8	.6333		8	1	.0060
	80	4	.6444		12	4	.0107
	81	6	.6611		16	5	.0167
	83	4	.6722		20	8	.0262
	84	4	.6833		24	2	.0286
	85	6	.7000		28	16	.0476
	86	8	.7222		32	10	.0595
	87	4	.7333		36	12	.0738
	88	4	.7444		40	4	.0786
	90	2	.7500		44	12	.0929
	91	4	.7611		48	12	.1071
	92	8	.7833		52	27	.1393
	93	4	.7944		56	6	.1464
	94	2	.8000		60	18	.1679
	97	2	.8056		64	21	.1929
	98	8	.8278		68	33	.2321
	100	4	.8389		72	15	.2500
	102	8	.8611		76	30	.2857
	104	4	.8722		80	21	.3107
	106	2	.8778		84	24	.3393
	108	4	.8889		88	30	.3750
	110	6	.9056		92	42	.4250
	112	4	.9167		96	24	.4536
	114	4	.9278		100	30	.4893
	116	4	.9389		104	24	.5179
	118	4	.9500		108	18	.5393
	120	4	.9611		112	42	.5893
	122	6	.9778		116	27	.6214
	128	4	.9889		120	18	.6429
	130	2	.9944		124	24	.6714
	134	2	1.0000		128	36	.7143
					132	12	.7286
					136	36	.7714
					140	24	.8000
					144	24	.8286
					148	12	.8429
					152	36	.8857
					156	12	.9000
					160	12	.9143
					164	6	.9214
					168	18	.9429
					176	12	.9571
					180	6	.9643
					184	12	.9786
					192	12	.9929
					200	6	1.0000

Table 18. Distributions of R_{fv} on S_n/S_{n-k}, for k = 5.

	$2R^2_{fv}$	Frequency	Cumulative Probability		$2R^2_{fv}$	Frequency	Cumulative Probability
n=5:	0	1	.0083	n=6:	0	1	.0014
	4	4	.0417		4	5	.0083
	8	3	.0667		8	6	.0167
	12	6	.1167		12	9	.0292
	16	7	.1750		16	16	.0514
	20	6	.2250		20	12	.0681
	24	4	.2583		24	14	.0875
	28	10	.3417		28	24	.1208
	32	6	.3917		32	20	.1486
	36	10	.4750		36	21	.1778
	40	6	.5250		40	23	.2097
	44	10	.6083		44	28	.2486
	48	6	.6583		48	24	.2819
	52	10	.7417		52	34	.3292
	56	4	.7750		56	20	.3569
	60	6	.8250		60	32	.4014
	64	7	.8833		64	42	.4597
	68	6	.9333		68	29	.5000
	72	3	.9583		72	29	.5403
	76	4	.9917		76	42	.5986
	80	1	1.0000		80	32	.6431
					84	20	.6708
					88	34	.7181
					92	24	.7514
					96	28	.7903
					100	23	.8222
					104	21	.8514
					108	20	.8792
					112	24	.9125
					116	14	.9319
					120	12	.9486
					124	16	.9708
					128	9	.9833
					132	6	.9917
					136	5	.9986
					140	1	1.0000

Table 19. Distributions of F^* on S_n/S_{n-k}, for k = 2.

	F^*	Frequency	Cumulative Probability		F^*	Frequency	Cumulative Probability
n=2:	0	1	.5000	n=11:	0	1	.0091
	2	1	1.0000		2	1	.0182
					18	9	.1000
n=3:	0	1	.1667		20	27	.3455
	2	2	.5000		36	72	1.0000
	4	3	1.0000				
				n=12:	0	1	.0076
n=4:	0	1	.0833		2	1	.0152
	2	1	.1667		20	10	.0909
	4	2	.3333		22	30	.3182
	6	6	.8333		40	90	1.0000
	8	2	1.0000				
				n=13:	0	1	.0064
n=5:	0	1	.0500		2	1	.0128
	2	1	.1000		22	11	.0833
	6	3	.2500		24	33	.2949
	8	9	.7000		44	110	1.0000
	12	6	1.0000				
				n=14:	0	1	.0055
n=6:	0	1	.0333		2	1	.0110
	2	1	.0667		24	12	.0769
	8	4	.2000		46	36	.2747
	10	12	.6000		48	132	1.0000
	16	12	1.0000				
				n=15:	0	1	.0048
n=7:	0	1	.0238		2	1	.0095
	2	1	.0476		26	13	.0714
	10	5	.1667		28	39	.2571
	12	15	.5238		52	156	1.0000
	20	20	1.0000				
n=8:	0	1	.0179				
	2	1	.0357				
	12	6	.1429				
	14	18	.4643				
	24	30	1.0000				
n=9:	0	1	.0139				
	2	1	.0278				
	14	7	.1250				
	16	21	.4167				
	28	42	1.0000				
n=10:	0	1	.0111				
	2	1	.0222				
	16	8	.1111				
	18	24	.3778				
	32	56	1.0000				

Table 20. Distributions of F^* on S_n/S_{n-k}, for k = 3.

	F^*	Frequency	Cumulative Probability
n=3:	0	1	.1667
	2	2	.5000
	4	3	1.0000
n=4:	0	1	.0417
	2	3	.1667
	4	7	.4583
	6	9	.8333
	8	4	1.0000
n=5:	0	1	.0167
	2	2	.0500
	4	5	.1333
	6	8	.2667
	8	20	.6000
	10	14	.8333
	12	10	1.0000
n=6:	0	1	.0083
	2	2	.0250
	4	3	.0500
	6	3	.0750
	8	12	.1750
	10	27	.4000
	12	18	.5500
	14	18	.7000
	16	30	.9500
	18	6	1.0000
n=7:	0	1	.0048
	2	2	.0143
	4	3	.0286
	8	4	.0476
	10	16	.1238
	12	36	.2952
	14	16	.3714
	16	12	.4286
	18	36	.6000
	20	60	.8857
	24	24	1.0000

	F^*	Frequency	Cumulative Probability
n=8:	0	1	.0030
	2	2	.0089
	4	3	.0179
	10	5	.0327
	12	20	.0923
	14	45	.2262
	16	20	.2857
	20	20	.3452
	22	60	.5238
	24	100	.8214
	30	60	1.0000
n=9:	0	1	.0020
	2	2	.0060
	4	3	.0119
	12	6	.0238
	14	24	.0714
	16	54	.1786
	18	24	.2262
	24	30	.2857
	26	90	.4643
	28	150	.7619
	36	120	1.0000
n=10:	0	1	.0014
	2	2	.0042
	4	3	.0083
	14	7	.0181
	16	28	.0569
	18	63	.1444
	20	28	.1833
	28	42	.2417
	30	126	.4167
	32	210	.7083
	42	210	1.0000
n=11:	0	1	.0010
	2	2	.0030
	4	3	.0061
	16	8	.0141
	18	32	.0465
	20	72	.1192
	22	32	.1515
	32	56	.2081
	34	168	.3778
	36	280	.6606
	48	336	1.0000

Table 21. Distributions of F^* on S_n/S_{n-k}, for k = 4.

n	F^*	Frequency	Cumulative Probability
n=4:	0	1	.0417
	2	3	.1667
	4	7	.4853
	6	9	.8333
	8	4	1.0000
n=5:	0	1	.0083
	2	4	.0417
	4	12	.1417
	6	24	.3417
	8	35	.6333
	10	24	.8333
	12	20	1.0000
n=6:	0	1	.0028
	2	3	.0111
	4	9	.0361
	6	19	.0889
	8	36	.1889
	10	70	.3833
	12	76	.5944
	14	78	.8111
	16	50	.9500
	18	18	1.0000
n=7:	0	1	.0012
	2	3	.0048
	4	7	.0131
	6	12	.0274
	8	19	.0500
	10	45	.1036
	12	99	.2214
	14	96	.3357
	16	144	.5071
	18	120	.6500
	20	168	.8500
	22	84	.9500
	24	42	1.0000
n=8:	0	1	.0006
	2	3	.0024
	4	7	.0065
	6	9	.0119
	8	8	.0167
	10	20	.0286
	12	60	.0643
	14	124	.1381
	16	108	.2024
	18	128	.2786
	20	168	.3786
	22	227	.5143
	24	324	.7071
	26	180	.8143
	28	120	.8857
	30	168	.9857
	32	24	1.0000
n=9:	0	1	.0003
	2	3	.0013
	4	7	.0036
	6	9	.0066
	8	4	.0079
	10	5	.0096
	12	25	.0179
	14	75	.0427
	16	155	.0939
	18	120	.1336
	20	120	.1733
	22	80	.1997
	24	280	.2923
	26	380	.4180
	28	500	.5833
	30	240	.6627
	32	180	.7222
	34	300	.8214
	36	420	.9603
	40	120	1.0000

Table 22. Distributions of F^* on S_n/S_{n-k}, for k = 5.

	F^*	Frequency	Cumulative Probability
n=5:	0	1	.0083
	2	4	.0417
	4	12	.1417
	6	24	.3417
	8	35	.6333
	10	24	.8333
	12	20	1.0000
n=6:	0	1	.0014
	2	5	.0083
	4	18	.0333
	6	46	.0972
	8	93	.2264
	10	137	.4167
	12	148	.6222
	14	136	.8111
	16	100	.9500
	18	26	1.0000
n=7:	0	1	.0004
	2	4	.0020
	4	14	.0075
	6	36	.0218
	8	81	.0540
	10	150	.1135
	12	286	.2270
	14	358	.3690
	16	460	.5516
	18	448	.7294
	20	376	.8786
	22	180	.9500
	24	126	1.0000

Table 23. Distributions of F_{fv} on S_n/S_{n-k}, for k = 2.

	F_{fv}	Frequency	Cumulative Probability
n=2:	0	1	.5000
	2	1	1.0000
n=3:	0	1	.1667
	2	2	.5000
	4	3	1.0000
n=4:	0	1	.0833
	2	1	.1667
	3	2	.3333
	5	6	.8333
	8	2	1.0000
n=5:	0	1	.0500
	2	1	.1000
	4	3	.2500
	6	9	.7000
	10	6	1.0000
n=6:	0	1	.0333
	2	1	.0667
	5	4	.2000
	7	12	.6000
	12	12	1.0000
n=7:	0	1	.0238
	2	1	.0476
	6	5	.1667
	8	15	.5238
	14	20	1.0000
n=8:	0	1	.0179
	2	1	.0357
	7	6	.1429
	9	18	.4643
	16	30	1.0000
n=9:	0	1	.0139
	2	1	.0278
	8	7	.1250
	10	21	.4167
	18	42	1.0000
n=10:	0	1	.0111
	2	1	.0222
	9	8	.1111
	11	24	.3778
	20	56	1.0000
n=11:	0	1	.0091
	2	1	.0182
	10	9	.1000
	12	27	.3455
	22	72	1.0000
n=12:	0	1	.0076
	2	1	.0152
	11	10	.0909
	13	30	.3182
	24	90	1.0000
n=13:	0	1	.0064
	2	1	.0128
	12	11	.0833
	14	33	.2949
	26	110	1.0000
n=14:	0	1	.0055
	2	1	.0110
	13	12	.0769
	15	36	.2747
	28	132	1.0000
n=15:	0	1	.0048
	2	1	.0095
	14	13	.0714
	16	39	.2571
	30	156	1.0000

Table 24. Distributions of F_{fv} on S_n/S_{n-k}, for k = 3.

	F_{fv}	Frequency	Cumulative Probability
n=3:	0	1	.1667
	2	2	.5000
	4	3	1.0000
n=4:	0	1	.0417
	2	3	.1667
	4	7	.4583
	6	9	.8333
	8	4	1.0000
n=5:	0	1	.0167
	2	2	.0500
	3	2	.0833
	4	3	.1333
	5	8	.2667
	7	18	.5667
	8	2	.6000
	9	8	.7333
	10	6	.8333
	12	10	1.0000
n=6:	0	1	.0083
	2	2	.0250
	4	6	.0750
	6	12	.1750
	8	27	.4000
	10	18	.5500
	12	18	.7000
	14	30	.9500
	18	6	1.0000
n=7:	0	1	.0048
	2	2	.0143
	4	3	.0286
	5	4	.0476
	7	16	.1238
	9	36	.2952
	11	16	.3714
	12	12	.4286
	14	36	.6000
	16	60	.8857
	21	24	1.0000
n=8:	0	1	.0030
	2	2	.0089
	4	3	.0179
	6	5	.0327
	8	20	.0923
	10	45	.2262
	12	20	.2857
	14	20	.3452
	16	60	.5238
	18	100	.8214
	24	60	1.0000
n=9:	0	1	.0020
	2	2	.0060
	4	3	.0119
	7	6	.0238
	9	24	.0714
	11	54	.1786
	13	24	.2262
	16	30	.2857
	18	90	.4643
	20	150	.7619
	27	120	1.0000
n=10:	0	1	.0014
	2	2	.0042
	4	3	.0083
	8	7	.0181
	10	28	.0569
	12	63	.1444
	14	28	.1833
	18	42	.2417
	20	126	.4167
	22	210	.7083
	30	210	1.0000
n=11:	0	1	.0010
	2	2	.0030
	4	3	.0061
	9	8	.0141
	11	32	.0465
	13	72	.1192
	15	32	.1515
	20	56	.2081
	22	168	.3778
	24	280	.6606
	33	336	1.0000

Table 25. Distributions of F_{fv} on S_n/S_{n-k}, for k = 4.

	F_{fv}	Frequency	Cumulative Probability
n=4:	0	1	.0417
	2	3	.1667
	4	7	.4853
	6	9	.8333
	8	4	1.0000
n=5:	0	1	.0083
	2	4	.0417
	4	12	.1417
	6	24	.3417
	8	35	.6333
	10	24	.8333
	12	20	1.0000
n=6:	0	1	.0028
	2	3	.0111
	3	2	.0167
	4	7	.0361
	5	10	.0639
	6	9	.0889
	7	30	.1722
	8	6	.1889
	9	62	.3611
	10	8	.3833
	11	48	.5167
	12	28	.5944
	13	40	.7056
	14	38	.8111
	16	50	.9500
	18	18	1.0000
n=7:	0	1	.0012
	2	3	.0048
	4	10	.0167
	6	24	.0452
	8	49	.1036
	10	99	.2214
	12	96	.3357
	14	144	.5071
	16	114	.6429
	18	156	.8286
	20	72	.9143
	22	30	.9500
	24	42	1.0000
n=8:	0	1	.0006
	2	3	.0024
	4	7	.0065
	5	4	.0089
	6	9	.0143
	7	20	.0262
	8	4	.0286
	9	60	.0643
	11	124	.1381
	12	12	.1482
	13	96	.2024
	14	48	.2310
	15	80	.2786
	16	168	.3786
	18	228	.5143
	20	300	.6929
	21	24	.7071
	22	108	.7714
	23	72	.8143
	25	120	.8857
	27	168	.9857
	32	24	1.0000
n=9:	0	1	.0003
	2	3	.0013
	4	7	.0036
	6	14	.0083
	8	29	.0179
	10	75	.0427
	12	155	.0939
	14	140	.1402
	16	180	.1997
	18	280	.2923
	20	380	.4180
	22	500	.5833
	24	240	.6627
	26	180	.7222
	28	300	.8214
	30	420	.9603
	36	120	1.0000

Table 26. Distributions of F_{fv} on S_n/S_{n-k}, for k = 5.

	F_{fv}	Frequency	Cumulative Probability
n=5:	0	1	.0083
	2	4	.0417
	4	12	.1417
	6	24	.3417
	8	35	.6333
	10	24	.8333
	12	20	1.0000
n=6:	0	1	.0014
	2	5	.0083
	4	18	.0333
	6	46	.0972
	8	93	.2264
	10	137	.4167
	12	148	.6222
	14	136	.8111
	16	100	.9500
	18	36	1.0000
n=7:	0	1	.0004
	2	4	.0020
	3	2	.0028
	4	12	.0075
	5	12	.0123
	6	24	.0218
	7	44	.0393
	8	37	.0540
	9	116	.1000
	10	34	.1135
	11	226	.2032
	12	60	.2270
	13	256	.3286
	14	102	.3690
	15	272	.4770
	16	188	.5516
	17	200	.6310
	18	248	.7294
	19	72	.7579
	20	304	.8786
	22	180	.9500
	24	126	1.0000

Table 27. Distributions of C^* on S_n/S_{n-k}, for k = 2.

	C^*	Frequency	Cumulative Probability		C^*	Frequency	Cumulative Probability
n=2:	0	1	.5000	n=14:	0	1	.0055
	1	1	1.0000		1	25	.1429
					2	156	1.0000
n=3:	0	1	.1667				
	1	3	.6667	n=15:	0	1	.0048
	2	2	1.0000		1	27	.1333
					2	182	1.0000
n=4:	0	1	.0833				
	1	5	.5000				
	2	6	1.0000				
n=5:	0	1	.0500				
	1	7	.4000				
	2	12	1.0000				
n=6:	0	1	.0333				
	1	9	.3333				
	2	20	1.0000				
n=7:	0	1	.0238				
	1	11	.2857				
	2	30	1.0000				
n=8:	0	1	.0179				
	1	13	.2500				
	2	42	1.0000				
n=9:	0	1	.0139				
	1	15	.2222				
	2	56	1.0000				
n=10:	0	1	.0111				
	1	17	.2000				
	2	72	1.0000				
n=11:	0	1	.0091				
	1	19	.1818				
	2	90	1.0000				
n=12:	0	1	.0076				
	1	21	.1667				
	2	110	1.0000				
n=13:	0	1	.0064				
	1	23	.1538				
	2	132	1.0000				

Table 28. Distributions of C^* on S_n/S_{n-k}, for k = 3.

	C^*	Frequency	Cumulative Probability
n=3:	0	1	.1667
	1	3	.6667
	2	2	1.0000
n=4:	0	1	.0417
	1	6	.2917
	2	11	.7500
	3	6	1.0000
n=5:	0	1	.0167
	1	9	.1667
	2	26	.6000
	3	24	1.0000
n=6:	0	1	.0083
	1	12	.1083
	2	47	.5000
	3	60	1.0000
n=7:	0	1	.0048
	1	15	.0762
	2	74	.4286
	3	120	1.0000
n=8:	0	1	.0030
	1	18	.0565
	2	107	.3750
	3	210	1.0000
n=9:	0	1	.0020
	1	21	.0437
	2	146	.3333
	3	336	1.0000
n=10:	0	1	.0014
	1	24	.0347
	2	191	.3000
	3	504	1.0000
n=11:	0	1	.0010
	1	27	.0283
	2	242	.2727
	3	720	1.0000
n=12:	0	1	.0008
	1	30	.0235
	2	299	.2500
	3	990	1.0000
n=13:	0	1	.0006
	1	33	.0198
	2	362	.2308
	3	1320	1.0000
n=14:	0	1	.0005
	1	36	.0169
	2	431	.2143
	3	1716	1.0000
n=15:	0	1	.0004
	1	39	.0147
	2	506	.2000
	3	2184	1.0000

Table 29. Distributions of C^* on S_n/S_{n-k}, for k = 4.

	C^*	Frequency	Cumulative Probability		C^*	Frequency	Cumulative Probability
n=4:	0	1	.0417	n=12:	0	1	.0001
	1	6	.2917		1	38	.0033
	2	11	.7500		2	539	.0487
	3	6	1.0000		3	3382	.3333
					4	7920	1.0000
n=5:	0	1	.0083				
	1	10	.0917	n=13:	0	1	.0001
	2	35	.3833		1	42	.0025
	3	50	.8000		2	659	.0409
	4	24	1.0000		3	4578	.3077
					4	11880	1.0000
n=6:	0	1	.0028				
	1	14	.0417				
	2	71	.2389				
	3	154	.6667				
	4	120	1.0000				
n=7:	0	1	.0012				
	1	18	.0226				
	2	119	.1643				
	3	342	.5714				
	4	360	1.0000				
n=8:	0	1	.0006				
	1	22	.0137				
	2	179	.1202				
	3	638	.5000				
	4	840	1.0000				
n=9:	0	1	.0003				
	1	26	.0089				
	2	251	.0919				
	3	1066	.4444				
	4	1680	1.0000				
n=10:	0	1	.0002				
	1	30	.0062				
	2	335	.0726				
	3	1650	.4000				
	4	3024	1.0000				
n=11:	0	1	.0001				
	1	34	.0044				
	2	431	.0588				
	3	2414	.3636				
	4	5040	1.0000				

Table 30. Distributions of C^* on S_n/S_{n-k}, for k = 5.

	C^*	Frequency	Cumulative Probability
n=5:	0	1	.0083
	1	10	.0917
	2	35	.3833
	3	50	.8000
	4	24	1.0000
n=6:	0	1	.0014
	1	15	.0222
	2	85	.1403
	3	225	.4528
	4	274	.8333
	5	120	1.0000
n=7:	0	1	.0004
	1	20	.0083
	2	155	.0698
	3	580	.3000
	4	1044	.7143
	5	720	1.0000

Table 31. Distributions of H^* on S_n/S_{n-k}, for k = 2.

	H^*	Frequency	Cumulative Probability
n=2:	0	1	.5000
	2	1	1.0000
n=3:	0	1	.1667
	2	3	.6667
	3	2	1.0000
n=4:	0	1	.0833
	2	5	.5000
	3	4	.8333
	4	2	1.0000
n=5:	0	1	.0500
	2	7	.4000
	3	6	.7000
	4	6	1.0000
n=6:	0	1	.0333
	2	9	.3333
	3	8	.6000
	4	12	1.0000
n=7:	0	1	.0238
	2	11	.2857
	3	10	.5238
	4	20	1.0000
n=8:	0	1	.0179
	2	13	.2500
	3	12	.4643
	4	30	1.0000
n=9:	0	1	.0139
	2	15	.2222
	3	14	.4167
	4	42	1.0000
n=10:	0	1	.0111
	2	17	.2000
	3	16	.3778
	4	56	1.0000
n=11:	0	1	.0091
	2	19	.1818
	3	18	.3455
	4	72	1.0000
n=12:	0	1	.0076
	2	21	.1667
	3	20	.3182
	4	90	1.0000
n=13:	0	1	.0064
	2	23	.1538
	3	22	.2949
	4	110	1.0000
n=14:	0	1	.0055
	2	25	.1429
	3	24	.2747
	4	132	1.0000
n=15:	0	1	.0048
	2	27	.1333
	3	26	.2571
	4	156	1.0000

Table 32. Distributions of H^* on S_n/S_{n-k}, for k = 3.

	H^*	Frequency	Cumulative Probability
n=3:	0	1	.1667
	2	3	.6667
	3	2	1.0000
n=4:	0	1	.0417
	2	6	.2917
	3	8	.6250
	4	9	1.0000
n=5:	0	1	.0167
	2	9	.1667
	3	14	.4000
	4	24	.8000
	5	12	1.0000
n=6:	0	1	.0083
	2	12	.1083
	3	20	.2750
	4	45	.6500
	5	36	.9500
	6	6	1.0000
n=7:	0	1	.0048
	2	15	.0762
	3	26	.2000
	4	72	.5429
	5	72	.8856
	6	24	1.0000
n=8:	0	1	.0030
	2	18	.0565
	3	32	.1518
	4	105	.4643
	5	120	.8214
	6	60	1.0000
n=9:	0	1	.0020
	2	21	.0437
	3	38	.1190
	4	144	.4048
	5	180	.7619
	6	120	1.0000
n=10:	0	1	.0014
	2	24	.0347
	3	44	.0958
	4	189	.3583
	5	252	.7083
	6	210	1.0000
n=11:	0	1	.0010
	2	27	.0283
	3	50	.0788
	4	240	.3212
	5	336	.6606
	6	336	1.0000
n=12:	0	1	.0008
	2	30	.0235
	3	56	.0659
	4	297	.2909
	5	432	.6182
	6	504	1.0000
n=13:	0	1	.0006
	2	33	.0198
	3	62	.0559
	4	360	.2657
	5	540	.5804
	6	720	1.0000
n=14:	0	1	.0005
	2	36	.0169
	3	68	.0481
	4	429	.2445
	5	660	.5467
	6	990	1.0000
n=15:	0	1	.0004
	2	39	.0147
	3	74	.0418
	4	504	.2264
	5	792	.5165
	6	1320	1.0000

Table 33. Distributions of H^* on S_n/S_{n-k}, for k = 4.

	H^*	Frequency	Cumulative Probability
n=4:	0	1	.0417
	2	6	.2917
	3	8	.6250
	4	9	1.0000
n=5:	0	1	.0083
	2	10	.0917
	3	20	.2583
	4	45	.6333
	5	44	1.0000
n=6:	0	1	.0028
	2	14	.0417
	3	32	.1306
	4	93	.3889
	5	136	.7667
	6	84	1.0000
n=7:	0	1	.0012
	2	18	.0226
	3	44	.0750
	4	153	.2571
	5	276	.5857
	6	276	.9143
	7	72	1.0000
n=8:	0	1	.0006
	2	22	.0137
	3	56	.0470
	4	225	.1810
	5	464	.4571
	6	600	.8143
	7	288	.9857
	8	24	1.0000
n=9:	0	1	.0003
	2	26	.0089
	3	68	.0314
	4	309	.1336
	5	700	.3651
	6	1080	.7222
	7	720	.9603
	8	120	1.0000
n=10:	0	1	.0002
	2	30	.0062
	3	80	.0220
	4	405	.1024
	5	984	.2976
	6	1740	.6429
	7	1440	.9286
	8	360	1.0000
n=11:	0	1	.0001
	2	34	.0044
	3	92	.0160
	4	513	.0808
	5	1316	.2470
	6	2604	.5758
	7	2520	.8939
	8	840	1.0000
n=12:	0	1	.0001
	2	38	.0033
	3	104	.0120
	4	633	.0653
	5	1696	.2081
	6	3696	.5192
	7	4032	.8586
	8	1680	1.0000
n=13:	0	1	.0001
	2	42	.0025
	3	116	.0093
	4	765	.0538
	5	2124	.1776
	6	5040	.4713
	7	6048	.8238
	8	3024	1.0000

Table 34. Distributions of H^* on S_n/S_{n-k}, for $k = 5$.

	H^*	Frequency	Cumulative Probability
n=5:	0	1	.0083
	2	10	.0917
	3	20	.2583
	4	45	.6333
	5	44	1.0000
n=6:	0	1	.0014
	2	15	.0222
	3	40	.0778
	4	135	.2653
	5	264	.6319
	6	265	1.0000
n=7:	0	1	.0004
	2	20	.0083
	3	60	.0321
	4	245	.1294
	5	604	.3690
	6	950	.7460
	7	640	1.0000

Table 35. Distributions of U^* on S_n/S_{n-k}, for k = 2.

	U^*	Frequency	Cumulative Probability
n=2:	0	1	.5000
	1	1	1.0000
n=3:	0	1	.1667
	1	4	.8333
	2	1	1.0000
n=4:	0	1	.0833
	1	3	.3333
	2	8	1.0000
n=5:	0	1	.0500
	1	4	.2500
	2	15	1.0000
n=6:	0	1	.0333
	1	5	.2000
	2	24	1.0000
n=7:	0	1	.0238
	1	6	.1667
	2	35	1.0000
n=8:	0	1	.0179
	1	7	.1429
	2	48	1.0000
n=9:	0	1	.0139
	1	8	.1250
	2	63	1.0000
n=10:	0	1	.0111
	1	9	.1111
	2	80	1.0000
n=11:	0	1	.0091
	1	10	.1000
	2	99	1.0000
n=12:	0	1	.0076
	1	11	.0909
	2	120	1.0000
n=13:	0	1	.0064
	1	12	.0833
	2	143	1.0000
n=14:	0	1	.0055
	1	13	.0769
	2	168	1.0000
n=15:	0	1	.0048
	1	14	.0714
	2	195	1.0000

Table 36. Distributions of U^* on S_n/S_{n-k}, for k = 3.

	U^*	Frequency	Cumulative Probability
n=3:	0	1	.1667
	1	4	.8333
	2	1	1.0000
n=4:	0	1	.0417
	1	9	.4167
	2	13	.9583
	3	1	1.0000
n=5:	0	1	.0167
	1	6	.1167
	2	27	.5667
	3	26	1.0000
n=6:	0	1	.0083
	1	7	.0667
	2	43	.4250
	3	69	1.0000
n=7:	0	1	.0048
	1	8	.0429
	2	61	.3333
	3	140	1.0000
n=8:	0	1	.0030
	1	9	.0298
	2	81	.2708
	3	140	1.0000
n=9:	0	1	.0020
	1	10	.0218
	2	103	.2262
	3	390	1.0000
n=10:	0	1	.0014
	1	11	.0167
	2	127	.1931
	3	581	1.0000
n=11:	0	1	.0010
	1	12	.0131
	2	153	.1677
	3	824	1.0000
n=12:	0	1	.0008
	1	12	.0106
	2	181	.1477
	3	1125	1.0000
n=13:	0	1	.0006
	1	14	.0087
	2	211	.1317
	3	1490	1.0000
n=14:	0	1	.0005
	1	15	.0073
	2	243	.1186
	3	1925	1.0000
n=15:	0	1	.0004
	1	16	.0062
	2	277	.1077
	3	2436	1.0000

Table 37. Distributions of U^* on S_n/S_{n-k}, for k = 4.

	U^*	Frequency	Cumulative Probability
n=4:	0	1	.0417
	1	9	.4167
	2	13	.9583
	3	1	1.0000
n=5:	0	1	.0083
	1	16	.1417
	2	61	.6500
	3	41	.9917
	4	1	1.0000
n=6:	0	1	.0028
	1	11	.0333
	2	63	.2083
	3	201	.7667
	4	84	1.0000
n=7:	0	1	.0012
	1	12	.0155
	2	91	.1238
	3	343	.5321
	4	393	1.0000
n=8:	0	1	.0006
	1	13	.0083
	2	121	.0804
	3	569	.4190
	4	976	1.0000
n=9:	0	1	.0003
	1	13	.0050
	2	153	.0556
	3	861	.3403
	4	1995	1.0000
n=10:	0	1	.0002
	1	15	.0032
	2	187	.0403
	3	1225	.2833
	4	3612	1.0000
n=11:	0	1	.0001
	1	16	.0021
	2	223	.0303
	3	1667	.2408
	4	6013	1.0000
n=12:	0	1	.0001
	1	17	.0015
	2	261	.0235
	3	2193	.2081
	4	9408	1.0000
n=13:	0	1	.0001
	1	18	.0011
	2	301	.0186
	3	2809	.1823
	4	14031	1.0000

Table 38. Distributions of U^* on S_n/S_{n-k}, for k = 5.

	U^*	Frequency	Cumulative Probability
n=5:	0	1	.0083
	1	16	.1417
	2	61	.6500
	3	41	.9917
	4	1	1.0000
n=6:	0	1	.0014
	1	25	.0361
	2	181	.2875
	3	381	.8167
	4	131	.9986
	5	1	1.0000
n=7:	0	1	.0004
	1	18	.0075
	2	143	.0643
	3	741	.3583
	4	1341	.8905
	5	276	1.0000

APPENDIX E - COMPARISON OF EXACT AND ASYMPTOTIC DISTRIBUTIONS

In Chapter V, it was shown that the metrics F_{fv}, R^2_{fv}, and H^* on S_n/S_{n-k} are asymptotically normally distributed, provided k/n is bounded away from 0 and 1 as n approaches infinity. Meanwhile, in Appendix D, the exact distributions of these metrics were tabulated, for various values of k and n.

The purpose of this appendix is to compare these exact and asymptotic distributions, to get some idea of the validity of the normal approximation. In particular, for the case n = 7 and k = 4, we will use some so-called "Q-Q plots" to compare the exact and asymptotic distributions of the metrics.

A Q-Q plot for comparing two probability distributions is a collection of points (x,y). The x-coordinate of each point is a certain quantile of the first distribution, while the y-coordinate of that point is the same quantile of the second distribution. The idea behind the Q-Q plot is that, if the two distributions are close together, then y should be approximately equal to x, and the collection of points (x,y) should lie close to the line y = x. Q-Q plots are discussed in greater detail in Gnanadesikan's book [G1].

Figures 6, 7, and 8 (pages 207-209) are the Q-Q plots of the exact distributions versus the asymptotic distributions of the metrics F_{fv}, R^2_{fv}, and H^* on S_n/S_{n-k}, where n = 7 and k = 4.

For illustrative purposes, we now describe how the first point in the lower left hand corner of Figure 6 was constructed. Table 25 on page 191 gives the exact distribution of F_{fv}. By Table 25 (with a couple of decimal places added), the value 0 is the p-th quantile of this exact distribution, for any p such that $0 < p \leq .001190$. We can choose p to be the average of the two extremes, and state that 0 is the .000595 quantile of the exact distribution of F_{fv}. On the other hand, the asymptotic theory of Chapter V implies that F_{fv} should be approximately normally distributed, with mean = 14.8571 and standard deviation = 4.6217. The .000595 quantile of this approximating normal distribution is -0.1225. Hence, the point $(x = 0, y = -0.1225)$ lies on the Q-Q plot. The remaining points in Figures 6, 7, and 8 were constructed similarly.

Most of the points in Figures 6, 7, and 8 lie very close to the line $y = x$. In Figures 7 and 8, it is true that there is some deviation from this line for certain extreme values of x, but this is to be expected, since we are approximating a bounded probability distribution by an unbounded one. Moreover, Table 17 (page 183) and Table 33 (page 199) show that these extreme points represent less than three percent of the total mass of these distributions.

In conclusion, the normal approximation is quite good in these examples. The applicability of the asymptotic theory is even more striking in light of the fact that the value $n = 7$ is not particularly large.

Figure 6. Q-Q Plot of the Exact Distribution versus the Asymptotic Distribution of the Metric F_{fv} on S_n/S_{n-k}, where $n = 7$ and $k = 4$.

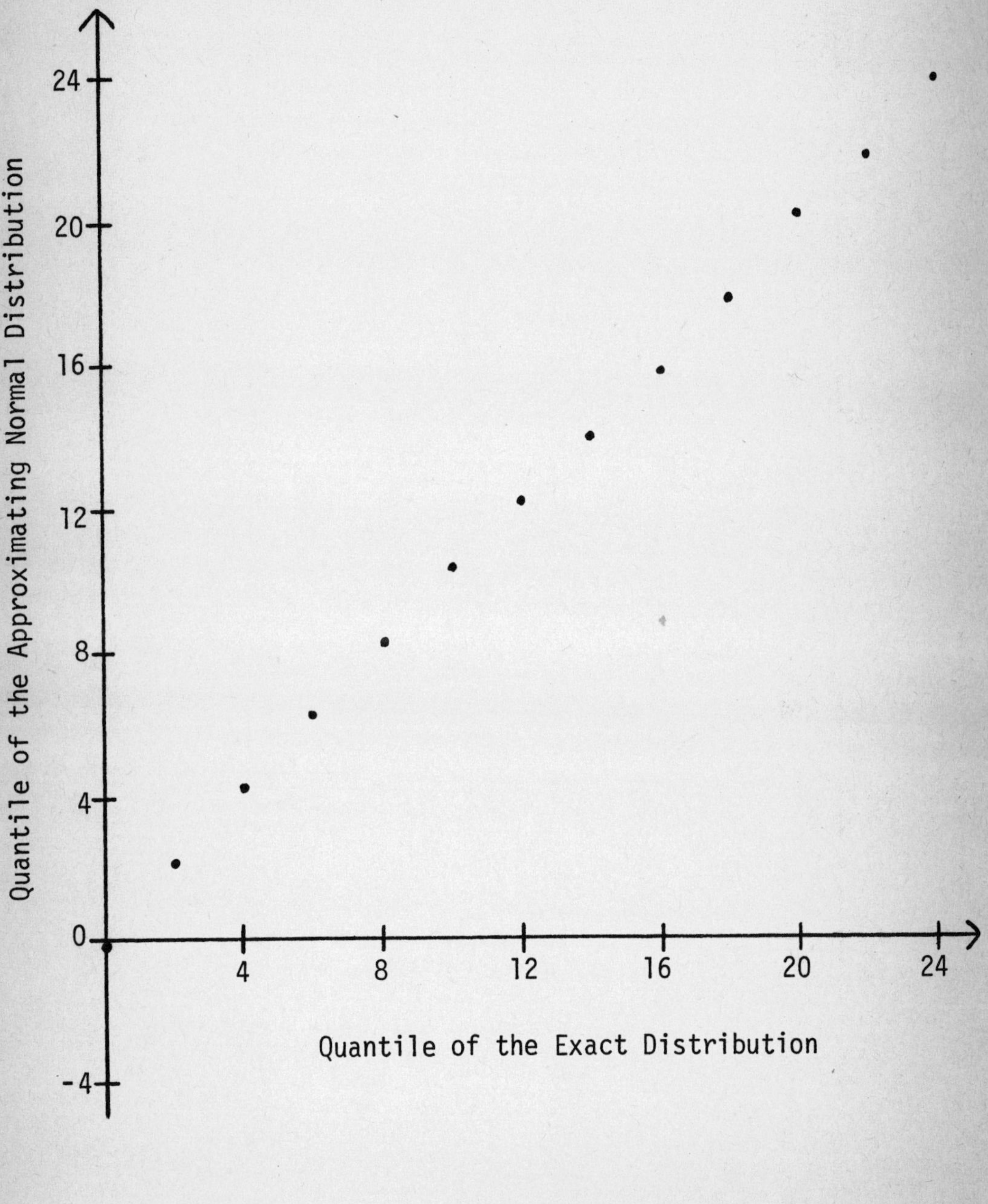

Figure 7. Q-Q Plot of the Exact Distribution versus the Asymptotic Distribution of R^2_{fv} on S_n/S_{n-k}, where $n = 7$ and $k = 4$.

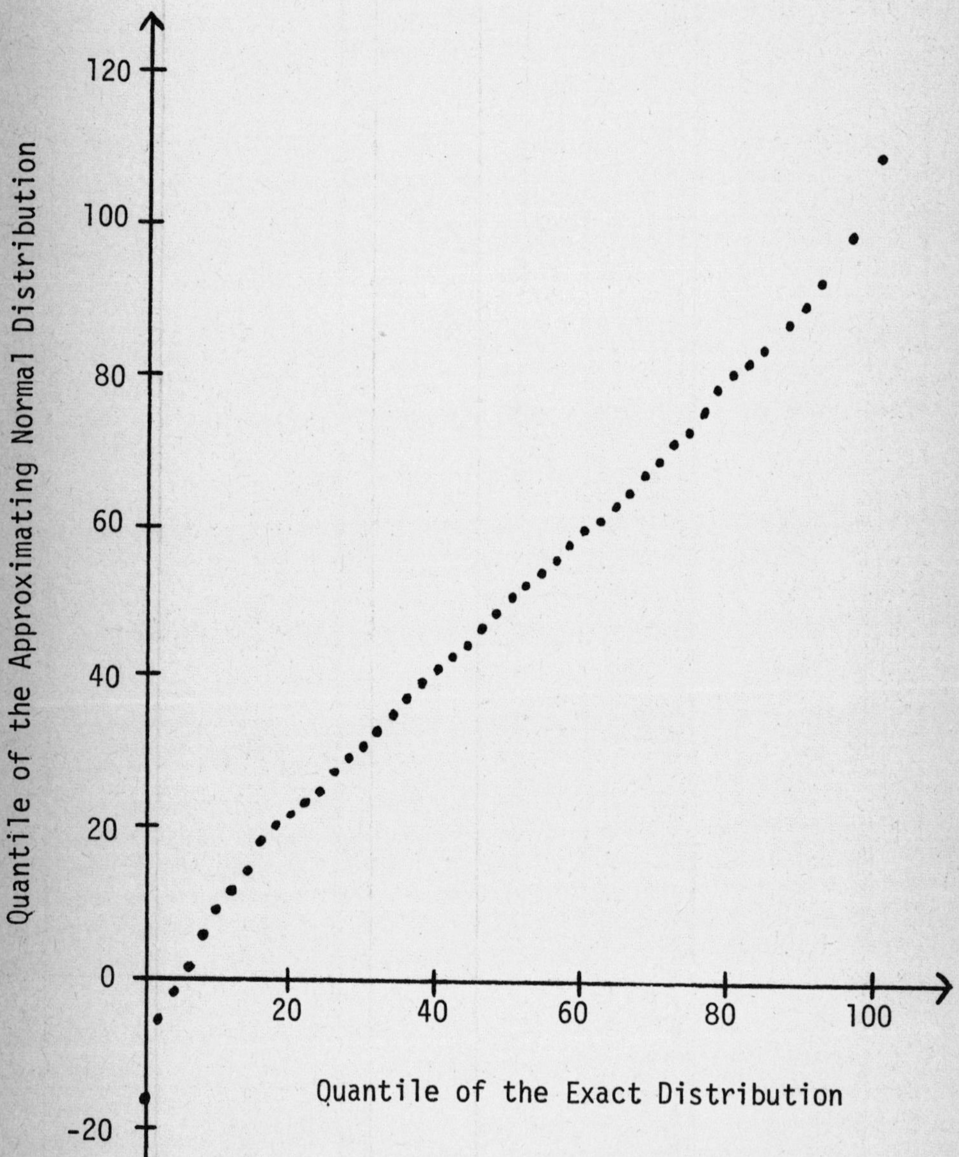

Figure 8. Q-Q Plot of the Exact Distribution versus the Asymptotic Distribution of the Metric H^* on S_n/S_{n-k}, where $n = 7$ and $k = 4$.

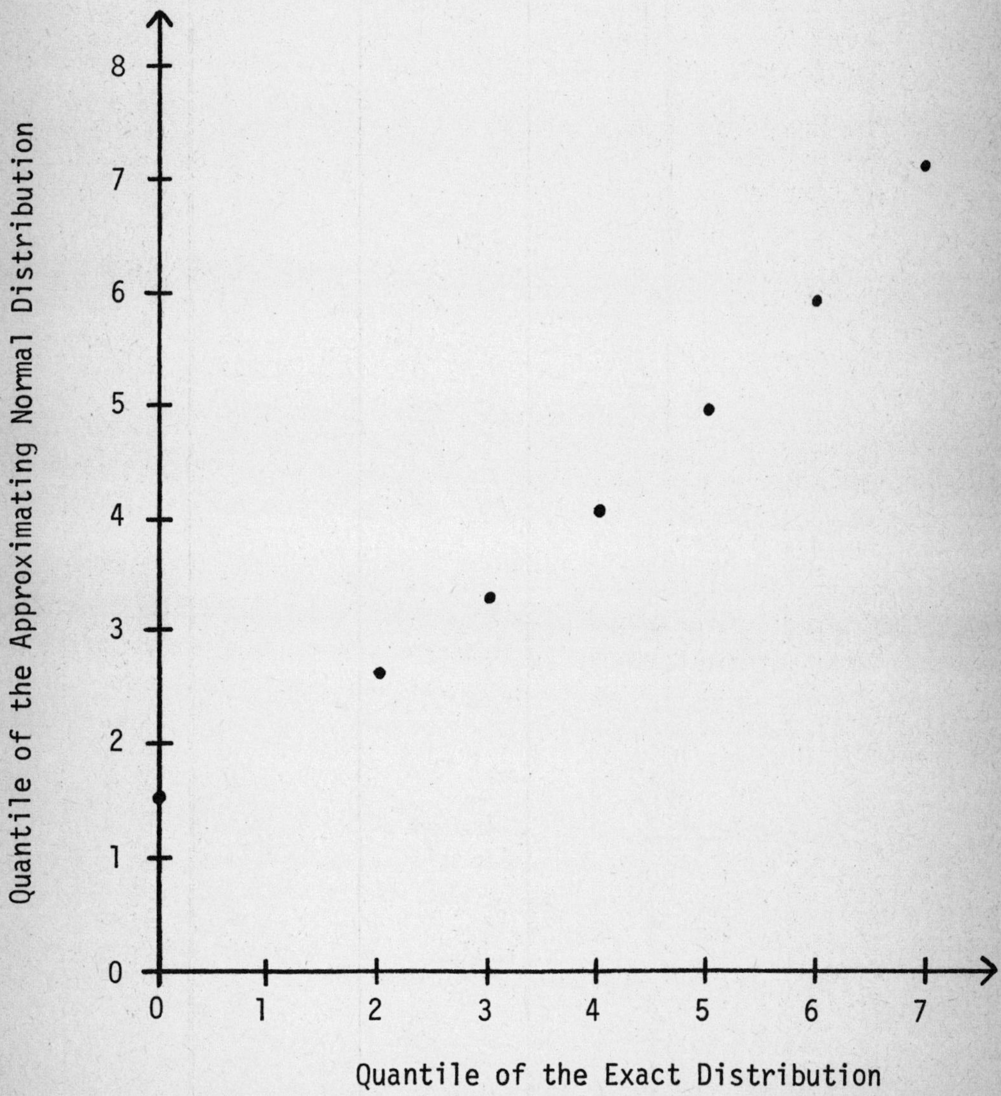

BIBLIOGRAPHY

[B1] Berlekamp, E. R. (1968). Algebraic Coding Theory. New York: McGraw-Hill Book Company.

[B2] Beyer, W. A., Stein, M. L., Smith, T. F., and Ulam, S. M. (1974). A Molecular Sequence Metric and Evolutionary Trees. Mathematical Biosciences, 19, pp. 9-25.

[B3] Birch, L. L. (1980). Effect of Peer Models' Food Choices and Eating Behaviors on Preschoolers' Food Preferences. Child Development, 51, pp. 489-496.

[B4] Blood, R. O. and Wolfe, D. M. (1960). Husbands and Wives. New York: Free Press.

[C1] Cameron, P. J. and Van Lint, J. H. (1980). Graphs, Codes, and Designs. London Mathematical Society Lecture Notes Series, Volume 43. Cambridge University Press.

[C2] Cayley, A. (1849). A Note on the Theory of Permutations. Philosophical Magazine, 34, pp. 527-529.

[D1] Davis, J. A. (1978). General Social Surveys, 1972-1978: Cumulative Codebook. Chicago: National Opinion Research Center.

[D2] Diaconis, P. (1982). Group Theory in Statistics. Harvard University Lecture Notes. To appear in the Institute of Mathematical Statistics Lecture Notes — Monograph Series.

[D3] Diaconis, P. and Graham, R. L. (1977). Spearman's Footrule as a Measure of Disarray. Journal of the Royal Statistical Society. Series B, Volume 39, pp. 262-268.

[D4] Dieudonné, J. (1969). Treatise on Analysis. Volumes I-VI. New York: Academic Press.

[F1] Feigin, P. and Cohen, A. (1978). On a Model for Concordance between Judges. Journal of the Royal Statistical Society, Series B, Volume 40, pp. 203-213.

[F2] Feller, W. (1968). An Introduction to Probability Theory and its Applications. Volume 1. Third Edition. New York: John Wiley and Sons.

[F3] Friedman, J. H. and Rafsky, L. C. (1979). Multivariate Generalizations of the Wald-Wolfowitz and Smirnov Two-Sample Tests. Annals of Statistics, 7, pp. 697-717.

[F4] Friedman, J. H. and Rafsky, L. C. (1983). Graph-Theoretic Measures of Multivariate Association and Prediction. Annals of Statistics, 11, pp. 377-391.

[G1] Gnanadesikan, R. (1977). Methods for Statistical Data Analysis of Multivariate Observations. New York: John Wiley and Sons.

[G2] Gordon, A. D. (1979). A Measure of the Agreement between Rankings. Biometrika, 66, pp. 7-15.

[G3] Green, P. E. and Carmone, F. J. (1970). Multidimensional Scaling and Related Techniques in Marketing Analysis. Boston: Allyn and Bacon.

[H1] Hausdorff, F. (1927). Mengenlehre. Third Edition. Berlin: W. de Gruyter and Company.

[H2] Herstein, I. N. (1964). Topics in Algebra. Lexington, Massachusetts: Xerox College Publishing.

[H3] Hoeffding, W. (1951). A Combinatorial Central Limit Theorem. Annals of Mathematical Statistics, 22, pp. 558-566.

[K1] Kelley, J. L. (1942). Hyperspaces of a Continuum. Transactions of the American Mathematical Society, 52, pp. 22-36.

[K2] Kendall, M. G. (1950). Discussion on Symposium on Ranking Methods. Journal of the Royal Statistical Society, Series B, Volume 12, p. 189.

[K3] Kendall, M. G. (1962). Rank Correlation Methods. Third Edition. New York: Hafner Publishing Company.

[K4] Kruskal, J. B. (1964). Multidimensional Scaling by Optimizing Goodness-of-Fit to a Nonmetric Hypothesis. Psychometrika, 29, pp. 1-27.

[K5] Kruskal, J. B. (1964). Nonmetric Multidimensional Scaling: A Numerical Method. Psychometrika, 29, pp. 115-129.

[K6] Kruskal, W. H. (1958). Ordinal Measures of Association. Journal of the American Statistical Association, 53, pp. 814-861.

[K7] Kuratowski, K. (1966). Topology: Volume I. New York: Academic Press.

[K8] Kuratowski, K. (1968). Topology: Volume II. New York: Academic Press.

[L1] Lancaster, H. O. (1969). The Chi-Squared Distribution. New York: John Wiley and Sons.

[L2] Logan, B. F. and Shepp, L. (1977). Variational Problems for Random Young Tableau. Advances in Mathematics, 26, pp. 206-222.

[M1] Mallows, C. L. (1957). Non-Null Ranking Models I. Biometrika, 44, pp. 114-130.

[M2] Mardia, K. V., Kent, J. B., and Bibby, J. M. (1979). Multivariate Analysis. London: Academic Press.

[M3] Michael, E. (1951). Topologies on Spaces of Subsets. Transactions of the American Mathematical Society, 71, pp. 152-182.

[N1] Nadler, S. B. (1978). Hyperspaces of Sets. New York: Marcel Dekker.

[P1] Pike, L. W. (1980). Implicit Guessing Strategies of Graduate Record Examination-Aptitude Examinees Classified by Ethnic Group and Sex. Graduate Record Examination Board Professional Report Number 75-10P. Princeton, New Jersey: Educational Testing Service.

[R1] Rukhin, A. L. (1972). Estimation of a Rotation Parameter on a Sphere. Zapiski Nauchnykh Seminarov LOMI, Volume 29 (1972), pp. 74-91 (in Russian). English Translation: Journal of Soviet Mathematics, Volume 3 (1975), pp. 777-791.

[S1] Sellers, P. H. (1974). On the Theory and Computation of Evolutionary Distances. SIAM Journal on Applied Mathematics, 26, pp. 787-793.

[S2] Serre, J.-P. (1977). Linear Representations of Finite Groups. New York: Springer-Verlag.

[S3] Silverberg, A. R. (1980). Statistical Models for Q-Permutations. Ph.D. Thesis, Department of Statistics, Princeton Unviersity.

[S4] Spearman, C. (1904). The Proof and Measurement of Association between Two Things. American Journal of Psychology, 15, pp. 72-101.

[S5] Spearman, C. (1906). A Footrule for Measuring Correlation. British Journal of Psychology, 2, pp. 89-108.

[U1] Ulam, S. M. (1972). Some Ideas and Prospects in Biomathematics. Annual Review of Biophysics and Bioengineering, 1, pp. 277-292.

[U2] Ulam, S. M. (1981). Future Applications of Mathematics in the Natural Sciences. American Mathematical Heritage: Algebra and Applied Mathematics. Texas Tech. University, Mathematics Series, Number 13, pp. 101-114.

[V1] Verducci, J. S. (1982). Discriminating between Two Populations on the Basis of Ranked Preferences. Ph.D. Thesis, Department of Statistics, Stanford University.

[W1] Waterman, M. S., Smith, T. F., and Beyer, W. A. (1976). Some Biological Sequence Metrics, Advances in Mathematics, 20, pp. 367-387.

INDEX OF NOTATION

INDEX OF NOTATION

(Continued)

INDEX OF NOTATION

(Continued)